WITHDRAWN

AXIOMATIC GEOMETRY

MICHAEL C. GEMIGNANI
Smith College

ADDISON-WESLEY PUBLISHING COMPANY, INC
Reading, Massachusetts · Menlo Park, California · London · Don Mills, Ontario

This book is in the
ADDISON-WESLEY SERIES IN MATHEMATICS EDUCATION

Consulting Editors
JOHN G. HARVEY and THOMAS A. ROMBERG

To Margaret, and John and Mary

Preface

Geometry, for a long time one of the most widely studied branches of mathematics, was one of the most neglected in the first half of the twentieth century. There now seems to be some resurgence of interest in this rich and often strikingly beautiful field of study. This renewed interest is fortunate for geometry can serve as an admirable introduction to abstract mathematics as well as a bridge between different areas of mathematical study.

Much of geometry is intuitively appealing and lends itself well to motivation. In geometry the student can often "picture" the situation he is investigating. The axioms for the geometries studied in Part I of this text are readily understood and easily assented to; *yet*, the conclusions which follow from these axioms not only are of great power and aesthetic appeal, but they often jar the intuition and make the student aware that even in mathematical systems whose axioms are few and easy to understand, the conclusions can be far-reaching and surprising, and can spill over into new areas of mathematics.

The treatment of this text is axiomatic. In Part I elementary geometries —linear spaces, affine and projective planes, and projective spaces—are developed. The student is led up to and through the algebraic implications of Desargues' Theorem concerning the collineations and coordinatization of affine and projective planes.

In Part II one of the classical axiomatizations of Euclidean plane geometry is explored, using modern terminology. Parts I and II together are intended to give the student a broad base in geometry and some familiarity with two quite dissimilar axiomatic developments. The parts are sufficiently in-dependent that Part II can be studied without Part I. It is recommended that the student have some introduction to linear algebra before undertaking this text.

I want especially to thank Professor Francis Bueckenhout for several useful conversations and papers which helped me in the design of Part I.

Northampton, Mass.　　　　　　　　　　　　　　　　　　　　　　　M.G.
November 1970

Contents

0   Preliminaries

0.1 SETS, FUNCTIONS, RELATIONS

It is assumed that any reader of this book has already had some experience
with sets; hence much of what is said in this section will be for the sake of
review rather than for the purpose of presenting new material.

A *set* will be taken to be any well-defined collection of objects; the objects
in a set are called *elements*, or *points*, of the set. If x is an element of the
set S, we write $x \in S$. We denote the phrase *is not an element of* by $\notin$.

Sets may be denoted either by explicitly listing their elements inside of
braces (for example, $\{1, 2, 3\}$ is the set having 1, 2, and 3 as elements) or by
giving the rule by which a typical object of the set is determined (for example,
$\{x \mid x$ is a red schoolhouse$\}$ is the set of all red schoolhouses, or, alternatively,
the set of all x such that x is a red schoolhouse).

A set S is said to be a *subset* of a set T if each element of S is an element
of T. We usually denote S *is a subset of* T by $S \subset T$. Two sets S and T
are *equal* if they contain exactly the same elements; that is, $S = T$ if $S \subset T$
and $T \subset S$. The phrase *is not a subset of* is denoted by $\not\subset$.

The empty set, that is, the set which contains no elements whatsoever, is
denoted by $\varnothing$.

If S and T are any two sets, then the *complement of S in T* is the set of
all elements of T which are not elements of S; we denote the complement of
S in T by $T - S$. Similarly, the *complement of T in S*, denoted by $S - T$,
is the set of all elements of S which are not elements of T.

The two most basic set operations are *union* and *intersection*. If $\{S_i\}$, $i \in I$, is any family of sets indexed by some set I^*, then the union of this family of sets is $\{x \mid x \in S_i$ for at least one $i \in I\}$. The union of $\{S_i\}$, $i \in I$, may be denoted by $\bigcup_I S_i$, or $\bigcup \{S_i \mid i \in I\}$. The *intersection* of this family of sets is

$$\{x \mid x \in S_i \text{ for each } i \in I, \text{ that is, } x \text{ is an element of} \\ \text{every member of the family of sets}\}.$$

The intersection of $\{S_i\}$, $i \in I$, may be denoted by $\bigcap_I S_i$, or $\bigcap \{S_i \mid i \in I\}$. Where only a few sets are involved, say $\{S_1, S_2, S_3\}$, the intersection and union of these sets may be denoted by $S_1 \cap S_2 \cap S_3$ and $S_1 \cup S_2 \cup S_3$, respectively. It is assumed that the reader is moderately familiar with these set operations, at least so far as any finite family of sets is concerned.

If S and T are any two sets, then the *Cartesian product* of S and T is defined to be the set of all ordered pairs (s, t) such that $s \in S$ and $t \in T$. The Cartesian product of S and T is denoted by $S \times T$.

If S and T are any sets, then a subset R of $S \times T$ is said to be a *relation between S and T*. A subset of $S \times S$ is said to be a *relation on S*. If R is a relation between S and T, that is, if $R \subset S \times T$, then if $(s, t) \in R$, we may also write sRt, or say that s and t are R-related. Although, strictly speaking, a relation is a set, at times a phrase or symbol defining the relation will be used in place of the actual set. For example, although *is equal to* defines a relation of the collection of subsets of some set, we usually write simply $S = T$ if S and T are equal subsets, rather than explicitly refer to any relation.

A *function f* from a set S into a set T is a relation between S and T such that each element of S is f-related to one and only one element of T. If $(s, t) \in f$, then we may write $t = f(s)$. Functions are usually defined by giving a rule which enables us to find $f(s)$ whenever s is given. Again, rarely is explicit mention made of the fact that a function is a set. Functions are also called *maps*, or *mappings*.

If f is a function from S into T, then S is called the *domain* of f, T the *range* of f, and $\{t \in T \mid t = f(s)$ for some $s \in S\}$ the *image* of f. The image of f may be denoted by $f(S)$. The image of f is a subset of the range of f, and usually is a *proper subset* of the range, that is, it is contained in, but is not equal to, the range.

If f is a function from S into T, we may write $f: S \rightarrow T$. If $f(S) = T$, then f is said to be *onto*. If $f(s) = f(s')$ implies $s = s'$ for any $s, s' \in S$, then f is said to be *one-one*; that is, f is one-one if each element of T is the image of at most one element of S.

---

* The reader may consider I to be merely a set of labels distinguishing the various members of the family of sets.

Example 1 Let S and T be the set of real numbers. The function f defined by $f(x) = 2x$ is both one-one and onto; that is, each real number is the function value of one and only one real number, namely, $y/2$. The function $g(x) = e^x$ is one-one, but not onto since no negative number occurs as a function value of g. The function $h(x) = x^2$ is neither one-one nor onto.

Suppose $f: S \rightarrow T$. If $t \in T$, then

$$f^{-1}(t) = \{s \in S \mid f(s) = t\}.$$

If $U \subset T$, then

$$f^{-1}(U) = \{s \in S \mid f(s) \in U\}.$$

By f^{-1} we mean $\{(t, s) \mid (s, t) \in f\}$. Note that f^{-1} is a relation between T and S, called the *inverse relation of f*, and that it is a function from T to S if and only if f is both one-one and onto.

Example 2 Suppose f is the function from the set S of real numbers into S defined by $f(x) = 3x - 1$. Since f is both one-one and onto, its inverse is a function. Specifically, if y is any real number, $f^{-1}(y) = \frac{1}{3}(y + 1)$. By definition, $f^{-1}(y)$ is that real number x such that $f(x) = y$. Compute $f(f^{-1}(y))$ to confirm that this is y.

Suppose $f: S \rightarrow T$ and $g: T \rightarrow W$. Then $g \circ f$, the *composition of g with f*, is defined by

$$\{(s, w) \mid s \in S, w \in W, \text{ such that there is some } t \in T$$
$$\text{with } t = f(s) \text{ and } w = g(t)\}.$$

More simply, $(g \circ f)(s) = g(f(s))$ for each $s \in S$.

Example 3 Suppose f and g are functions from the set S of real numbers into S defined by $f(x) = x^2$ and $g(x) = 4x - 1$. Then

$$(f \circ g)(x) = f(g(x)) = f(4x - 1) = (4x - 1)^2,$$

while

$$(g \circ f)(x) = g(f(x)) = g(x^2) = 4x^2 - 1.$$

The reader is already familiar with the ordering of the real numbers by *less than or equal to*. If R is the set of real numbers and "less than or equal to" is denoted by $\leq$, then $\leq$ defines a relation on R having the properties that

P1 $x \leq x$, for any $x \in R$;

P2 $x \leq y$ and $y \leq x$ implies $x = y$, for any $x, y \in R$; and

P3 $x \leq y$ and $y \leq z$ implies $x \leq z$, for any x, y, and z in R.

Any relation on any set S which shares properties P1 through P3 is called a *partial ordering* on S. If S has a partial ordering defined on it, then S is said to be a *partially ordered set*. We may denote a set S with a partial ordering $\leq$ by $S, \leq$.

A partial ordering $\leq$ of a set S which satisfies the additional property

$$T) \qquad \text{given } s, s' \in S, \qquad s \leq s' \text{ or } s' \leq s,$$

is said to be a *total ordering* of S. The usual ordering of the real numbers is a total ordering. The example below gives a partial ordering which is not a total ordering.

Example 4 Let $P(S)$ be the family of subsets of a set S. Then $\subset$ defines a partial ordering on $P(S)$. We verify that $\subset$ satisfies P1 through P3.

P1 If W is any subset of S, then $W \subset W$.

P2 If W and T are two subsets of S such that $W \subset T$ and $T \subset W$, then $W = T$.

P3 If W, T, and Z are sets such that $W \subset T$ and $T \subset Z$, then each element of W is also an element of T. But since $T \subset Z$, each element of T is also an element of Z; therefore each element of W is an element of Z, that is, $W \subset Z$.

Given two subsets W and T of S, it is not necessarily true that $W \subset T$ or $T \subset W$; therefore $\subset$ is not a total ordering of $P(S)$.

Since the less than or equal to relation on the set of real numbers is the prototype of a partial ordering, we will generally denote a partial ordering by $\leq$, unless there is a special symbol called for.

Let S, $\leq$ be any partially ordered set, and suppose $W \subset S$. An element u of S is said to be an *upper bound* for W if $w \leq u$ for each $w \in W$. An element v of S is said to be a *lower bound* for W if $v \leq w$ for each $w \in W$. It is not necessarily true that every nonempty subset of a partially ordered set has an upper or a lower bound. An element U of S is said to be a *least upper bound* for W if U is an upper bound for W and $U \leq u$ for each upper bound u of W. We say that an element V of S is a *greatest lower bound* for W if V is a lower bound for W and $v \leq V$ for each lower bound v of W. The least upper bound and greatest lower bound of W may be denoted by lub W and glb W, respectively. It is not necessarily the case that W has either a lub or glb.

Example 5 Let Q be the set of rational numbers totally ordered by $\leq$. Let W be the set of rational numbers less than $\sqrt{2}$. Then 3 is an upper bound for W; but since $\sqrt{2}$ is an irrational number, it can be shown that W has no least upper bound in Q. Note that W does have a least upper bound in the full set of real numbers, namely, $\sqrt{2}$.

A partial ordering is an example of a special kind of relation that can be defined on a set. Another particularly important type of relation is an *equivalence relation*. The prototype for an equivalence relation is $=$, just as $\leq$ is the prototype for a partial ordering. Since ambiguity is likely to result if $=$ is used to denote an arbitrary equivalence relation, E will be used instead. A relation E on a set S is said to be an *equivalence relation* on S if E satisfies the following properties

E1 s E s for any $s \in S$.

E2 If s and s' are elements of S such that s E s', then s' E s.

E3 If s, s', and s'' are elements of S such that s E s' and s' E s'', then s E s''.

Compare E1 through E3 with the properties of $=$. Note too that the only difference between a partial ordering on S and an equivalence relation on S is that property P2 has been replaced by property E2.

Example 6 Let T be the set of all plane triangles. Then *is similar to, is congruent to*, and *has the same area as*, all define equivalence relations on T.

The most important property of an equivalence relation is given in the following proposition.

Proposition 1 *Let S be any set. A **partition** $\mathscr{P}$ of S is any collection of non-empty subsets of S such that each element of S is contained in one and only one member of $\mathscr{P}$. Suppose E is an equivalence relation on S. For each $s \in S$, set $\bar{s} = \{t \in S \mid s \text{ E } t\}$. Then the collection of $\bar{s}$ for all $s \in S$ is a partition of S, called the **partition induced by** E.*

Proof We want to show that each element of S is in one and only one member of $\{\bar{s}\}$, $s \in S$. Since a E a for each $a \in S$ by E1, then $a \in \bar{a}$; hence each element of S is contained in at least one member of $\{\bar{s}\}$, $s \in S$. Suppose that $a \in \bar{a}$ and $a \in \bar{t}$. Choose any $a' \in \bar{a}$. Then a E a'; also t E a since $a \in \bar{t}$. By E3, t E a and a E a' implies t E a'; hence $a' \in \bar{t}$. Therefore $\bar{a} \subset \bar{t}$. A similar argument, however, shows that $\bar{t} \subset \bar{a}$; hence $\bar{a} = \bar{t}$. Thus $\bar{a}$ is the only member of $\{\bar{s}\}$, $s \in S$, which contains a. Since a was an arbitrary element of S, $\{\bar{s}\}$, $s \in S$, is a partition of S.

If E is an equivalence relation on a set S, then if s E s', s and s' are said to be E-*equivalent*, or simply, *equivalent*. The set of elements of S which are equivalent to an element s of S (the set $\bar{s}$ in the proof above) is said to be the E-*equivalence class of s*, or simply the *equivalence class of s*.

EXERCISES

1. Express each of the following in words.

   a) $\{a, b, c, d\}$

   b) $\{a, b, \{c, d\}\}$

   c) $\{n \mid n \text{ is an integer and } 2 \text{ divides } n\}$

   d) $\{\varnothing\}$

   e) $a \in \{a, b, c\}$

   f) $a \notin \{\{a\}\}$

   g) $P \not\subset \{P\}$

2. Prove each of the following.
   a) If $S \subset T$, then $T - (T - S) = S$.
   b) If S is any set, then $\varnothing \subset S$.
   c) $S \subset T$ if and only if $S \cap T = S$.
   d) If $f: S \to T$ and $A \subset T$, then $f\big(f^{-1}(A)\big) \subset A$.
   e) If $f: S \to T$ and $A \cup B \subset T$, then $f^{-1}(A \cup B) = f^{-1}(A) \cup f^{-1}(B)$.
   f) If $f: S \to T$ and $\{U_i\}$, $i \in I$, is any family of subsets of T, then
   $$f^{-1}\big(\bigcap\{U_i \mid i \in I\}\big) = \bigcap\{f^{-1}(U_i) \mid i \in I\}.$$
   g) If $f: S \to T$ and $g: T \to W$, then $(g \circ f)^{-1}(A) = f^{-1}\big(g^{-1}(A)\big)$ for any $A \subset W$.
   h) If $f: S \to T$, then f^{-1} is a function from T into S if and only if f is one-one and onto.

3. Let $\mathscr{P}$ be a partition of a set S and define $s \text{ E } s'$ if and only if s and s' are contained in the same member of $\mathscr{P}$. Prove that E defines an equivalence relation on S, the equivalence classes of which are the members of $\mathscr{P}$.

4. Let N be the set of positive integers, and $n \mid m$ denote n *divides* m, that is, $m = nk$ for some positive integer k. Prove that $\mid$ defines a partial ordering on N. Show that this partial ordering of N is not a total ordering. Prove that any two-element subset of N, $\mid$ has both a least upper bound and a greatest lower bound. Find several subsets which have no upper bound (with respect to $\mid$). Is there a nonempty subset of N which has no lower bound with respect to $\mid$?

5. Prove that if U and U' are both lub's of W, $W \subset S$, $\le$, then $U = U'$.

0.2 CARDINALITY

Two sets S and T are said to have the *same number of elements*, or to have the *same cardinality*, if there is a one-one function from S onto T. That is, S and T have the same number of elements if the elements of S can be put into one-one correspondence with the elements of T.

A set S is said to be *finite* if S is empty, or if there is a positive integer n such that S has the same cardinality as $\{1, 2, \ldots, n\}$. Otherwise, S is said to be *infinite*. Furthermore, a set S is said to be *countable* if S has the same number of elements as a subset of N, the set of positive integers. Otherwise, S is said to be *uncountable*. Thus any finite set is certainly countable.

Throughout this chapter we will use N to denote the set of positive integers.

Proposition 2 *Any subset of a finite set is finite. Any subset of a countable set is countable.*

Proof We prove the first statement and leave the proof of the second as an exercise. Let S be a finite set. Since S is finite, either $S = \varnothing$, or there is a positive integer n such that S has the same number of elements as

$\{1, 2, \ldots, n\}$. If $S = \emptyset$, then the only subset of S is $\emptyset$, which is finite. Assume that $S \neq \emptyset$. Then there is a one-one function f from S onto $\{1, 2, \ldots, n\}$ for an appropriate n. Suppose $W \subset S$. If $W = \emptyset$, then W is finite. If $W \neq \emptyset$, let $i_1, i_2, \ldots, i_m$ be the elements of $\{1, 2, \ldots, n\}$ in the image of W. Then defining $g \colon W \to \{1, 2, \ldots, m\}$ by $g(w) = j$, where $f(w) = i_j$, for each $w \in W$, we see that W is finite.

Proposition 3 *Let* $\{A_n\}$, $n \in N$, *be a countable collection of countable sets. Then* $\bigcup_N A_n$ *is also countable* (N *represents the set of positive integers*).

Proof We may enumerate the elements of each of the A_n in an array as shown.

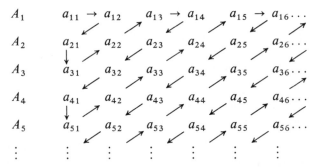

The element a_{nm} is the mth element of A_n. If we run out of elements in any set, that is, if any of these sets are finite, we just put down x's in the spot where an element should go.

We now must find a one-one function f from $\bigcup_N A_n$ onto some subset of the set N of positive integers. Set

$$f(a_{11}) = 1, \qquad f(a_{12}) = 2, \qquad \text{and} \qquad f(a_{21}) = 3.$$

In general, follow the path indicated in the diagram, deleting any repeated elements after they have been encountered once, and correspond the kth element reached with k. Eventually any element of $\bigcup_N A_n$ will be reached; hence $\bigcup_N A_n$ can be put in one-one correspondence with a subset of N, and is therefore countable.

Corollary 1 *If A and B are countable sets, then $A \times B$ is countable.*

Proof Let $A = \{a_1, a_2, a_3, \ldots\}$ and $B = \{b_1, b_2, b_3, \ldots\}$. Set

$$A_n = \{(a_n, b) \mid b \in B\}, \quad \text{for each } n \in N.$$

Then each A_n has the same number of elements as B, and hence is countable. Therefore $\bigcup \{A_n \mid n \in N\} = A \times B$ is countable by Proposition 3.

Corollary 2 *The set Q of rational numbers is countable.*

Proof If q is any positive rational number, then we may consider q as the quotient of two positive integers m/n, where m/n is a fraction in lowest terms. Associate q with the ordered pair (m, n). We see then that the positive rational numbers can be associated with a subset of $N \times N$, where N is the set of positive integers. But $N \times N$ is countable by Corollary 1. Therefore, by Proposition 2, the set of positive rational numbers is countable. The set of negative rational numbers, however, has the same number of elements as the set of positive rational numbers (corresponding q with $-q$), and hence the set of negative rational numbers is countable. But Q is the union of $\{0\}$, the set of positive rational numbers, and the set of negative rational numbers, all three of which are countable sets; therefore Proposition 3 tells us Q is countable.

Corollary 3 *The set Z of integers is countable.*

Proof The set Z is a subset of Q, and hence is countable by Proposition 2.

Although we now have a goodly number of sets we know to be countable, we have not yet shown that any set is uncountable. The following example shows that the set of real numbers is uncountable.

Example 7 The set S of unending decimals between 0 and 1 which contain only 0 or 1 as digits is uncountable. For proof, suppose that S is countable. Then we can find a one-one correspondence f between S and the set N of positive integers; hence we can make a table like the following, in which the first column gives a positive integer and the second column the element of S associated with it by the suitable function f.

n	$f(n)$
1	$0.1101100\ldots$
2	$0.0101000\ldots$
3	$0.1111101\ldots$
4	$0.1010101\ldots$
$\vdots$	$\vdots$

We now form an element $.x_1x_2x_3x_4\ldots$ of S as follows: If the first digit of $f(1)$ is 0, let $x_1 = 1$, and if the first digit of $f(1)$ is 1, let $x_1 = 0$. Similarly, if the second digit of $f(2)$ is 0, let $x_2 = 1$, and if it is 1, let $x_2 = 0$. In general, we let x_n be 0 or 1 depending on whether the nth digit of $f(n)$ is 1 or 0. Then $.x_1x_2x_3\ldots$ could not be $f(n)$ for any positive integer n, since $.x_1x_2x_3\ldots$ differs from each $f(n)$ at least in the nth digit because of the way it has been constructed. Hence f could not be onto, and S is therefore uncountable.

Now S is a subset of R, the set of real numbers. If R were countable, then, by Proposition 2, S would also be countable; therefore R is uncountable.

EXERCISES

1. Prove that any two infinite subsets of the set of positive integers have the same number of elements.

2. Prove that "has the same number of elements as" defines an equivalence relation on the collection of all sets.

3. Prove that no finite set S has the same number of elements as one of its proper subsets W. (A subset W of S is said to be *proper* if $W \neq S$.) Does this remain true if S is infinite?

4. Let $P(S)$ denote the set of subsets of the set S. Prove that S and $P(S)$ do not have the same number of elements.

5. Prove that any subset of a countable set is countable.

0.3 GROUPS, FIELDS, VECTOR SPACES

Although it is hoped that the reader already knows some group theory, such knowledge is not absolutely essential for this text, since in this section we will define those notions which the reader will have to know later.

Let S be any set. An *operation on S* is a function $\#$ from $S \times S$ into S. If $\#(s, s') = s''$, we usually write $s \# s' = s''$. Thus addition and multiplication are examples of operations on the set of real numbers.

An operation $\#$ on a set is said to be *associative* if for any elements s_1, s_2, and s_3 of S,

$$(s_1 \# s_2) \# s_3 = s_1 \# (s_2 \# s_3).$$

We thus recognize addition and multiplication of real numbers to be associative operations.

A set S with operation $\#$ may be denoted by $S, \#$. An element k of $S, \#$ is said to be an *identity with respect to $\#$* if

$$s \# k = k \# s = s$$

for any $s \in S$. Thus 0 is an identity with respect to addition of real numbers, and 1 is an identity with respect to multiplication of real numbers.

Suppose $S, \#$ has an identity k with respect to $\#$. An element t of S is said to be an *inverse* (with respect to $\#$) for an element s of S if

$$s \# t = t \# s = k.$$

If $\#$ is associative and t is an inverse for s, then it can be shown that t is the only inverse for s; in such a case, the unique inverse of s is denoted by s^{-1}.

If r is any real number, then the inverse of r with respect to $+$ is $-r$. If $r = 0$, then r has no inverse with respect to multiplication, but if $r \neq 0$, then the inverse of r with respect to multiplication is $r^{-1} = 1/r$.

A set S with operation $\#$ is said to be a *group* (with respect to $\#$) if (1) $\#$ is associative, (2) S has an identity k with respect to $\#$, and (3) each element of S has an inverse with respect to $\#$.

If $\#$ is such that $s \# s' = s' \# s$ for each s and s' in S, then $\#$ is said to be *commutative*. If S, $\#$ is a group and $\#$ is commutative, then S, $\#$ is called a *commutative group*, or an *abelian group*.

Suppose that S, $\#$ and T, $\$$ are groups. There may be many functions from S into T, but perhaps only a few of these are related in any way to the group structures (that is, to $\#$ and $\$$) of S and T. When studying groups, however, we wish to consider functions which somehow respect the operations of the groups; such functions are called homomorphisms. More formally, a function $f: S \rightarrow T$ is called a *homomorphism* if

$$f(s_1 \# s_2) = f(s_1) \, \$ \, f(s_2)$$

for any elements s_1 and s_2 of S. If f is a one-one and onto function as well as being a homomorphism, then f is said to be an *isomorphism*, and the groups S, $\#$ and G, $\$$ are said to be *isomorphic*. Isomorphic groups have essentially the same group properties.

If S, $\#$ and T, $\$$ are groups, then we can define an operation $\&$ on $S \times T$ as follows. If (s, t) and (s', t') are any elements of $S \times T$, define

$$(s, t) \, \& \, (s', t') = (s \# s', t \, \$ \, t').$$

We call the group $S \times T$, $\&$, thus formed, the *direct sum* of S, $\#$ and T, $\$$. We denote the direct sum of S, $\#$ and T, $\$$ by $S \oplus T$.

If G, $\#$ is a group and H is a subset of G such that H, $\#$ is also a group, then we call H a *subgroup* of G. A subgroup H of G is said to be a *normal subgroup* of G if, given any elements g of G and h of H, $g^{-1} \# h \# g$ is an element of H. Normal subgroups are very important in the study of groups, but, for the purposes of this text, it suffices simply to know the definition of a normal subgroup.

Let H be a subgroup of the group G, $\#$. For each element g of G, the sets

$$g \# H = \{g \# h \mid h \in H\} \qquad \text{and} \qquad H \# g = \{h \# g \mid h \in H\}$$

are called the *left* and *right cosets of g modulo H*, respectively. One can prove that H is a normal subgroup of G if and only if $g \# H = H \# g$ for each $g \in G$. The next proposition states a fundamental property of cosets.

Proposition 4 *Let G, $\#$ be any group and H a subgroup of G. Then $\{g \# H\}$, $g \in G$, is a partition of G; that is, each element of G is in one and only one coset modulo H.*

(The partition mentioned in the proposition is actually that induced by the equivalence relation defined by "g is equivalent to g' if and only if $g^{-1} \# g' \in H$".)

Some sets have two operations defined on them. For example, one can both add and multiply real numbers. A set S with operations $+$ and $\cdot$ is said to be a *field* if (1) S, $+$ is a commutative group, (2) $S - \{0\}, \cdot$, where 0 is the identity with respect to $+$, is a commutative group with respect to $\cdot$; and (3) $\cdot$ is *distributive* over $+$, that is, $s \cdot (s' + s'') = s \cdot s' + s \cdot s''$ and $(s' + s'') \cdot s = s' \cdot s + s'' \cdot s$ for any elements s, s', and s'' of S.

If $F, +, \cdot$ is a field, we will, for simplicity, call $+$ addition and $\cdot$ multiplication, and write $a \cdot b$ as ab provided that no ambiguity will result.

If F is a field and F' is a subset of F, which is a field in its own right with respect to the operations of F, then F' is said to be a *subfield* of F. The field of rational numbers is a subfield of the field of real numbers, which, in turn, is a subfield of the field of complex numbers. The following example gives an important family of fields.

Example 8 Let Z be the set of integers with the usual operations of addition and multiplication. Then Z is not a field (although it has many of the properties of a field) since not all integers have integer inverses with respect to multiplication. Let p be a *prime integer* (that is, $p \neq -1, 1$ and p is divisible only by 1, -1, p and $-p$). We define an integer n to be *congruent modulo p* to an integer m if n and m both leave the same remainder on division by p; we will denote *is congruent modulo p* by $\equiv \pmod{p}$. Then $\equiv \pmod{p}$ defines an equivalence relation on Z. There are p equivalence classes, one for each possible remainder on division by p; we denote these classes by $\bar{0}, \bar{1}, \ldots, \overline{p-1}$. Note that $\bar{0}$ contains all integers evenly divisible by p. The set of equivalence classes is denoted by Z_p and called the *set of integers modulo p*.

An algebraic structure can be put on Z_p as follows: To add $\bar{i}$ and $\bar{j}$, we set $\bar{i} + \bar{j} = \overline{i + j}$; for multiplication, set $\bar{i}\bar{j} = \overline{ij}$. It can be shown that Z_p with addition and multiplication so defined is a field containing p elements. The addition and multiplication tables for Z_3 are given below.

$+$	$\bar{0}$	$\bar{1}$	$\bar{2}$
$\bar{0}$	$\bar{0}$	$\bar{1}$	$\bar{2}$
$\bar{1}$	$\bar{1}$	$\bar{2}$	$\bar{0}$
$\bar{2}$	$\bar{2}$	$\bar{0}$	$\bar{1}$

$\cdot$	$\bar{0}$	$\bar{1}$	$\bar{2}$
$\bar{0}$	$\bar{0}$	$\bar{0}$	$\bar{0}$
$\bar{1}$	$\bar{0}$	$\bar{1}$	$\bar{2}$
$\bar{2}$	$\bar{0}$	$\bar{2}$	$\bar{1}$

Let V, $+$ be a commutative group and S be a set. We say that S is a set of *operators* on V if there is a map $\#$ from $S \times V$ into V; thus for each element v of V and s of S, $\#(s, v) = s \# v$ is an element of V. For example, let R, $+$ be the additive group of real numbers. Then R can be thought of as a set of operators on R, $+$ with $s \# r$ defined to be sr for any real numbers s and r. When S is a set of operators on a group V, $+$, we generally use sv to indicate the element of V obtained when s operates on v (that is, when we take $s \# v$).

A commutative group $V, +$ is said to be a *vector space over the field F* if (1) F is a set of operators on V, (2) $a(v + v') = av + av'$ for any $a \in F$ and v and v' in V, (3) $(a + a') v = av + a' v$ for any $a, a' \in F$ and $v \in V$, (4) $(ab) v = a(bv)$ for any $a, b \in F$ and $v \in V$, and (5) $1v = v$ for any $v \in V$, where 1 is the identity with respect to multiplication in F. The operation of an element of F on an element of V is generally referred to as *scalar multiplication*.

The following example gives an important family of vector spaces. Virtually no other type of vector space will be encountered in this text.

Example 9   Let F be any field and F^n be the set of all ordered n-tuples $(x_1, \ldots, x_n)$ whose coordinates are elements of F. We define

$$(a_1, \ldots, a_n) + (b_1, \ldots, b_n) = (a_1 + b_1, \ldots, a_n + b_n),$$

and

$$a(c_1, \ldots, c_n) = (ac_1, \ldots, ac_n);$$

that is, we define addition and scalar multiplication in F^n to be coordinatewise. Then it is easily shown that F^n is a vector space over F.

An element of a vector space is called a *vector*.

Let V be a vector space over the field F. A *linear combination* of vectors $v_1, \ldots, v_n$ is defined to be any sum of scalar multiples of the v_i, that is, a vector of the form $a_1 v_1 + a_2 v_2 + \cdots + a_n v_n$, where the a_i are elements of F. A set S of vectors is said to *generate* V if each vector in V can be represented as a linear combination of the vectors in S. The set S is said to be a *basis* for V if each vector in V can be expressed in one and only one way as a linear combination of the vectors in S.

Example 10   One can show that the vectors,

$$(1, 0, 0, \ldots, 0), (0, 1, 0, \ldots, 0), (0, 0, 1, 0, \ldots, 0), \ldots,$$
$$(0, 0, \ldots, 1, 0), (0, 0, \ldots, 0, 1)$$

(that is, the set of vectors with 1 in exactly one coordinate and 0 in the rest), form a basis for F^n. Whenever we refer to F^n in such a way as to imply a basis, we will mean this basis.

The number of elements in a basis for a vector space does not depend on which particular basis is used. The number of elements in any (and hence every) basis of a vector space is called the *dimension* of the vector space.

Continue to let V be a vector space over F. A subset W of V which is itself a vector space over F is said to be a *subspace* of V. A set of vectors in V is said to be *linearly independent* if it is the basis for some subspace of V. (Alternately, a subset S of V is linearly independent if no proper subset of S generates the same set of vectors that S does.)

It remains to deal briefly with functions between vector spaces. Suppose V and V' are both vector spaces over a field F. A function f from V into V' is said to be a *linear transformation* if

$$f(av + bv') = af(v) + bf(v')$$

for any $a, b \in F$ and $v, v' \in V$. That is, a linear transformation is one which respects both vector addition and scalar multiplication.

Since any vector of V has a unique expression as a linear combination of basis elements, a linear transformation is uniquely determined by how it acts on members of a basis of V. Specifically, if $v_1, \ldots, v_n$ is a basis for V and f is a linear transformation from V into V', then, given any vector v of V,

$$f(v) = f(a_1 v_1 + \cdots + a_n v_n) = a_1 f(v_1) + \cdots + a_n f(v_n),$$

where $a_1 v_1 + \cdots + a_n v_n$ is the unique representation of v in terms of the v_i.

Suppose now that f is a linear transformation from F^n into F^m. Let e_i be the vector (in either F^n or F^m) with 1 in the ith coordinate and 0's elsewhere. Then we find

$$f(e_1) = a_{11} e_1 + \cdots + a_{1m} e_m$$

$$f(e_2) = a_{21} e_1 + \cdots + a_{2m} e_m$$

$$\vdots$$

$$f(e_n) = a_{n1} e_1 + \cdots + a_{nm} e_m,$$

for suitable elements a_{ij}, $i = 1, \ldots, n; j = 1, \ldots, m$, of F. Since f is completely determined by what it does to $e_1, \ldots, e_m$, a basis of F^m, the a_{ij} completely define f. The array

(1)
$$\begin{pmatrix} a_{11} & a_{12} & \cdots & a_{1m} \\ a_{21} & a_{22} & \cdots & a_{2m} \\ \vdots & & & \\ a_{n1} & a_{n2} & \cdots & a_{nm} \end{pmatrix}$$

is called the *matrix* of f. The linear transformation f is completely determined by its matrix; moreover, any array such as (1) is the matrix of a unique linear transformation of V into V'.

This discussion of groups, fields, and vector spaces, although containing almost all that the student will have to know in order to read this text, is necessarily over-brief. Preferably, the student should have had some introduction to linear algebra before attempting this book. The following references may prove useful to the reader as further sources on the material found in this summary.

1. G. Birkhoff and S. MacLane, *A Survey of Modern Algebra*, New York; Macmillan, 1953.
2. W. Ledermann, *Introduction to the Theory of Finite Groups*, London; Oliver and Boyd, 1961.
3. R. Johnson, *Linear Algebra*, New York; Prindle, Weber, and Schmidt, 1967.

EXERCISES

1. Let S be any set. Prove that the set of one-one functions from S onto S is a group with composition as the group operation. Is this group necessarily commutative?

2. Prove that if H and H' are subgroups of G, $\#$ such that $H \cap H' = \{k\}$, where k is the identity of G and each element g of G can be expressed as $g = h \# h'$ for some element h of H and h' of H', then G is isomorphic to $H \oplus H'$.

3. Prove that F^n as given in Example 9 is a vector space over F. Prove the vectors claimed to form a basis for F^n in Example 10 do in fact form a basis.

4. Verify the addition and multiplication tables given for Z_3 in Example 5. Construct such tables for Z_5.

5. Suppose f is a homomorphism from the group G, $\#$ into the group H, \$. Prove that the image of f is a subgroup of H. Let k' be the identity of H. Prove that $f^{-1}(k')$ is a normal subgroup of G.

6. Define what is meant by an *isomorphism* from one vector space onto another. Prove that any two isomorphic vector spaces have the same dimension.

7. Prove that the intersection of any family of subspaces of some vector space is again a subspace. Suppose W and W' are subspaces of the vector space V. Define $W + W' = \{w + w' \mid w \in W \text{ and } w' \in W'\}$. Prove that $W + W'$ is a subspace of V.

8. Prove that the composition of any two homomorphisms is a homomorphism. Prove that the composition of two linear transformations is a linear transformation.

9. Prove that the direct sum of two groups is a group.

10. Prove Proposition 4.

Part I   Linear Geometry.
Affine and
Projective Planes

1 An Elementary Geometry

1.1 INTRODUCTION

The geometries presented in this text are of different degrees of complexity. The basic projective plane has a great deal less structure than the Euclidean plane; the latter requires a substantial number of axioms to define it. However, a projective plane and the Euclidean plane are both examples of geometries. The purpose of this chapter is to abstract from all geometries, both simple and complex, the basic properties which make them geometries. The purpose of this abstraction is threefold.

First, by abstracting the least common denominator of all geometries, we will come to understand better what a geometry is and what the study of geometry is all about. Second, by studying a basic geometry, we can see most clearly the general questions that arise in the study of geometries. Once the fundamental questions are posed in the case of the most elementary geometry, we can try to answer these questions in the case of a more complex geometry. Third, we wish to start this book as clearly and simply as possible, and build to more complicated geometric concepts and techniques of proof. We want to build this study of geometry like a house, from the foundation up.

Our method will be *axiomatic*. We will define geometries using undefined, or *primitive*, terms and *axioms*. The axioms tell us how the primitive terms are related to one another. When the primitive terms of a geometry can be assigned to the elements and notions associated with some "concrete" system S in such a way that the axioms of the geometry become true statements about S, then S is said to be a *model* for the geometry. We will give models for most of the geometries we study, but geometries can be studied "in the abstract" without reference to any "real" system.

The student must be wary of trying to impose on a geometry a statement he feels must be true about the geometry simply because the statement is true in some model of the geometry. The only statements which are true about a geometry are those which can be proved from the axioms for the geometry. A model for a geometry may give insights into the geometry as well as some idea of what is true about the geometry and how to go about proving it, but the truth of a statement in a model for a geometry does not necessarily mean that the statement follows from the axioms defining the geometry, much less does it constitute a proof for the statement.

It must be conceded that there is a study which does draw conclusions about certain geometries by studying models of the geometries; this study is called *analytic geometry*. (The axiomatic approach which we will use is known as *synthetic geometry*.) For example, the coordinate plane R^2 consisting of all ordered pairs (x, y) of real numbers is used to study Euclidean plane geometry analytically. Points are the elements of R^2, while lines are graphs of equations of the form $y = mx + b$ or $x = a$. Other geometric notions, such as *line segment* and *congruence*, pertinent to Euclidean plane geometry can be interpreted in the context of R^2. The algebraic properties of the real numbers and of R^2 are used to define geometric concepts and figures and to draw conclusions about Euclidean plane geometry.

Since we intend to be axiomatic in our approach, we will not go into a detailed exposition of analytic geometry. We should point out, though, that the contradiction between our statement that geometric truth depends on provability from axioms and the fact that truth in analytic geometry is established in the context of a model is only an apparent contradiction. Whether we are dealing with analytic or synthetic geometry, whatever is proved within the context of a particular model is true only about that model. Some models are of such great importance that the study of these models is significant and useful in its own right even if the results obtained happen only to be valid for those models.

However, in some instances, it can also be shown that all models for some geometry (as defined by a set of primitive terms and axioms) are identical, except possibly for the way things are labeled; that is, by suitably renaming the components of one model for the geometry, we can get any other model. In such an instance the geometry could be investigated by studying one of its models, since all models are basically the same in the way in which the basic notions of the geometry are related to one another. If all models of a geometry were "equivalent" in some suitable sense (to be discussed later in the text), we would not risk finding some statement about the geometry to be true in one model which could not be applied to any other model (once the terms of the statement had been properly translated into notions associated with the other model). Although we will in fact see examples of equivalent models later in this text, we illustrate the point we are trying to make in the following example.

Example 1 Let the axioms for a "geometry" be

Axiom 1 *Any two points are incident to precisely one line.*

Axiom 2 *Any two lines are incident to precisely one point.*

Axiom 3 *There are precisely three lines.*

We will call a system satisfying Axioms 1–3 a *semitriangular geometry*. The primitive notions of this geometry are *point, line,* and *incident to*. These terms are left undefined, but the axioms indicate how they are related to one another.

There are at least two dissimilar models for a semitriangular geometry. Let $S = \{1, 2, 3\}$. We will use S to construct two models for a semitriangular geometry.

Model 1 Let S itself be the only *point* and let each element of S be a *line*; interpret *are incident to* as meaning *determine*. With the interpretations of the primitive terms applied to S as indicated, S becomes a model for a semitriangular geometry. (The reader may find S to be a trivial and unsatisfying model of a semitriangular geometry. Nevertheless, this detracts in no way from its being a *bona fide* model of a semitriangular geometry. Personal feelings must sometimes be set aside in order to arrive at the facts.)

Model 2 Let each element of S be called a *point* and each pair of elements of S be called a *line*; again we interpret *are incident to* as meaning *determine*. Now there are three points and three lines instead of one point and three lines as in Model 1; Models 1 and 2 cannot be "equivalent" since Model 2 has more points than does Model 1. Figure 1.1 gives a pictorial representation of Model 2.

We now add another axiom.

Axiom 4 *There are precisely three points.*

We call the system defined by Axioms 1–4 a *triangular geometry*.

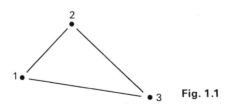

Fig. 1.1

Model 1 is an example of a semitriangular geometry which is not triangular, while Model 2 is both semitriangular and triangular. The following proposition can be proved about triangular geometries.

Proposition 1 *Each line is incident to exactly two points and each point is incident to exactly two lines.*

Proof We merely outline the proof and leave the details as an exercise. Label the three points of the geometry x, y, and z. Then the lines incident to x and y, y and z, and x and z are all distinct; hence these three lines are the only lines. For if some line were incident to more than two points, then some line would be incident to all three points, which leads to a contradiction of Axiom 3. From this it follows that each point is incident to precisely two lines. It also follows from this that each line is incident to precisely two points.

For distinct points x and y, we will denote the line incident to x and y by xy. Thus, the lines of a triangular geometry with points x, y, and z are xy, yz, and xz. Moreover, xy, yz, and xz are incident only to x and y, y and z, and x and z, respectively. It follows, then, that all triangular geometries look essentially like the one pictured in Fig. 1.1. The points might be labeled A, B, and C, or x, y, and z, instead of 1, 2, and 3, but that is immaterial. Figure 1.1 would be appropriate for any triangular geometry provided that the elements of the figure were suitably labeled. Since all triangular geometries are basically the same as that of Model 2, we can study what is true about points, lines, and incidence in Model 2 and feel certain that our conclusions will apply to all triangular geometries. We could not study Model 2 and be sure that whatever conclusions could be drawn from the model concerning the primitive terms would apply in all semitriangular geometries. For example, the statement, "There are three points," is true in Model 2, but not in Model 1.

Geometries, in general, deal among other things with lines and points. In the next section we let lines and points have the simplest possible relationship to one another consistent with the notion of a geometry and we investigate the consequences of that relationship.

EXERCISES

1. Supply the details for the proof of Proposition 1. In particular, prove each of the following:
   a) The lines xy, yz, and xz are distinct.
   b) Each point is incident to precisely two lines. Specifically, x is incident to xy and zx, etc.

c) Each line is incident to precisely two points. Specifically, xy is incident only to x and y, etc.

2. Supply a model of a semitriangular geometry which has precisely four points, or prove that any semitriangular geometry contains at most three points.

3. Consider the *lines* to be the lines of the coordinate plane R^2 with equations $y = x$, $y = -x$, and $y = 0$, and the *points* to be the points $(1, -1)$, $(0, 0)$, and $(1, 1)$ of R^2. Let *is incident to* be interpreted in the sense of *determines*. Which of the Axioms 1–4 does this system fail to satisfy?

Now let the above remain, except that the line with equation $y = 0$ is replaced by the line with equation $x = 1$ (Fig. 1.2). Does this new system satisfy Axioms 1–4? In other words, does this give a model of a triangular geometry? Does Fig. 1.2 look like Fig. 1.1? Is the statement, "Each line looks like the real line," true with regard to this system? Is the statement true with regard to all triangular geometries? Do we have any real contradiction here of the statement that all triangular geometries essentially look like Model 2 (from the viewpoint of their "triangular geometric" properties)? Explain carefully.

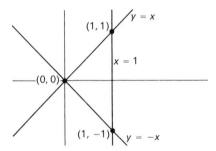

Fig. 1.2

4. A set of axioms is said to be *independent* if each axiom in the set cannot be proved or disproved from the other axioms in the set. Are Axioms 1–4 independent? Try to find a model which satisfies all of the axioms except for Axiom i for $i = 1, 2, 3$, and 4. Explain why having a model which satisfies all of the axioms except Axiom 1 proves that Axiom 1 cannot be proved or disproved from the other axioms.

5. We call a system defined by Axioms 1–3 and Axiom 4′ below a *quasi-triangular geometry*.

Axiom 4′ *Each point is incident to at least two lines.*

Prove or disprove: A geometry is triangular if and only if it is quasi-triangular. [*Hint:* Is Axiom 4′ a true theorem about triangular geometries? Is Axiom 4 a true theorem about quasi-triangular geometries? If the answer to both questions is yes, then we have triangular if and only if quasi-triangular. Explain why. What would be true if the answer were yes to only the first question? to only the second question?]

1.2 LINEAR SPACES

One normally thinks of a geometry as being a particular type of structure on a set. In particular, in a geometry we usually have a set S together with certain subsets of S which are the *points, lines, planes,* and other types of geometric objects*. Since we want to start simply, we will assume that there are but points and lines in a geometry. We expect two points to determine a line and two lines to intersect in a point if they intersect at all. We might have other properties that we feel should be true about every geometry, but the following axioms are certainly fundamental if a set S is to have a geometric structure where certain of its subsets are lines and points.

Axiom 1.1 *Any two distinct points x and y of S belong to (that is, are contained in, or are subsets of) one and only one line in S. We denote the unique line containing x and y by xy.*

Axiom 1.2 *Each line in S contains at least two points.*

Definition 1 *A set S having points and lines satisfying Axioms 1.1 and 1.2 is called a **linear space**.*

Point and *line* are the primitive terms in the axioms for a linear space. We have avoided the relationship *are incident to* (or its equivalent), which appears in the previous section, by explicitly considering the geometry of a linear space to be a structure on a set of points with a point "incident to" a line and the line incident to the point if and only if the point is contained in the line. We now give several examples of linear spaces.

Example 2 Let S be a set of at least two elements. The elements of S (or, more correctly, the one-element subsets of S) are the points†. Let the lines be all two-element subsets of S. Clearly, any two points are contained in a unique two-element subset of S, and each two-element subset of S contains at least two points. Thus S, with the structure described, is a linear space.

Example 3 Ordinary Euclidean plane geometry is a linear space. The set S consists of the points of the Euclidean plane. The lines are the usual lines of Euclidean geometry. This example must necessarily remain informal for the time being since we have only an informal idea at present of what Euclidean geometry is; a rigorous definition will be given later in the text.

---

* Assuming that there is an underlying set is not the only way to approach the study of geometry. For example, in the previous section the notion of a triangular geometry was introduced without reference to an underlying set. Nevertheless, we usually think of a geometry as referring to a structure on a point set, and the assumption of an underlying set obviates certain complications without really detracting from the generality of the study.

† In general, when the points are one-element subsets of S we will not distinguish between the points as subsets and the elements of S.

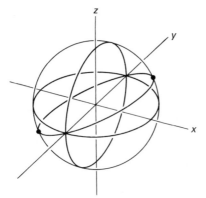

Fig. 1.3

Nevertheless, from the experience the reader has had with Euclidean plane geometry, he should readily recognize it as a linear space. (In fact this geometry is one of the prime inspirations for the axioms which define a linear space.)

Example 4 Let S be the sphere of radius 1 in Euclidean 3-space (Fig. 1.3). Let each point of S be a pair of antipodal elements of S. Then we have a linear space if we define a line in S to be a subset of S consisting of all of the points (pairs of antipodal elements) of S which lie on the same great circle of S. This geometry has the interesting property that any two lines must intersect (since any two great circles on the sphere must intersect in a pair of antipodal points).

Example 5 Suppose S has the structure of a linear space and S' is a subset of S which contains at least two points. Let each point of S contained in S' be a point of S'. If x and y are points of S', let the line in S' which contains x and y be defined as $xy \cap S'$ (where xy is the unique line in S which contains both x and y). Then S' has the structure of a linear space in its own right. (Why would it not suffice to say: Let the lines of S' be the intersections of S' with the lines of S?) The subset S' with the geometric structure thus defined is said to be the *linear space induced on S' by S*.

We will shortly prove some of the basic properties of linear spaces. However, since a linear space is defined by so few and such broad axioms, we cannot expect to be able to say very much which will apply to all linear spaces. That is, since what will apply to all linear spaces will apply to virtually all geometries, what can be said about linear spaces will necessarily be very general in nature. We could not, for example, expect to be able to say anything about *congruence* in general linear spaces since congruence is a notion proper to Euclidean geometry, and Euclidean geometry is a very specialized type of linear space. Nevertheless, there are interesting notions which can be introduced in the context of linear spaces and these notions point the way toward some of the things that might be of interest in less general geometries.

Proposition 2 *In any linear space, two distinct lines contain at most one point in common.*

Proof Suppose L and L' are lines of a linear space and both contain the points x and y. If x and y are distinct, then they are both contained in precisely one line xy. But then $L = xy = L'$.

Definition 2 *Let S be a linear space. A set of points of S are said to be **collinear** if they are all contained in some one line. If the points are not all contained in some line, they are said to be **noncollinear**.*

Euclidean solid geometry is a linear space. In this geometry, there are planes in addition to lines and points. A plane in solid geometry is determined by three noncollinear points, just as a line is determined by two points. The plane determined by the noncollinear points P, Q, and R can be defined entirely in terms of lines in the following way: The plane PQR containing P, Q, and R consists of all points on any line which contains at least two points of the set $PQ \cup QR \cup PR$ (Fig. 1.4). A plane can also be characterized in Euclidean solid geometry as a set which contains a line whenever it contains any two points of the line, and which contains at least two distinct lines but not all lines. We make no attempt to prove the assertions about Euclidean solid geometry made here. What we are trying to point out is that higher-dimensional geometric objects, such as planes, can be defined in terms of lines and points. We now generalize this observation to linear spaces.

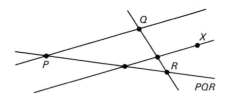

Fig. 1.4

Definition 3 *A subset V of a linear space S is said to be a **linear variety** if given distinct points x and y of V, the line xy is a subset of V.*

Thus, a plane is a linear variety in Euclidean solid geometry. If S is a linear space, then S, each line of S, and each point of S are linear varieties.

Proposition 3 *In a linear space S the intersection of any family of linear varieties of S is a linear variety.*

Proof Suppose that $\mathscr{V}$ is a family of linear varieties of S. Let x and y be distinct points of $\bigcap \{V \mid V \text{ in } \mathscr{V}\}$. Then x and y are also points in each V in $\mathscr{V}$. Therefore the line xy is a subset of each V in $\mathscr{V}$ since each V is a linear variety. Consequently, xy is a subset of $\bigcap \{V \mid V \text{ in } \mathscr{V}\}$. Hence, whenever x and y are distinct points of $\bigcap \{V \mid V \text{ in } \mathscr{V}\}$, xy is also in this intersection; therefore $\bigcap \{V \mid V \text{ in } \mathscr{V}\}$ is a linear variety.

Proposition 4 *Any subset T of a linear space S is a subset of at least one linear variety, namely, S itself.*

Corollary *If T is any subset of a linear space S, then there is a linear variety $V(T)$ which contains T and is a subset of every linear variety which contains T; that is, $V(T)$ is the smallest linear variety which contains T.*

Proof Let $V(T)$ be the intersection of all linear varieties which contain T. Because S contains T and is a linear variety, $V(T)$ is not the empty set; moreover, $V(T)$ is a linear variety which contains T. Now if W is any linear variety which contains T, then $V(T)$ is contained in the intersection of W and S, which is W; therefore $V(T)$ is a subset of W.

Definition 4 *For any subset T of a linear space S, the linear variety $V(T)$ described in the corollary to Proposition 4 is called the **linear closure** of T. We continue to denote the linear closure of T by $V(T)$.*

If V is a linear variety and T is a subset of S such that $V = V(T)$, then T is said to *generate* V.

Example 6 If T is a subset of a set S with the linear structure described in Example 2, then $V(T) = T$. Every subset T of S in this case is a linear variety; for if x and y are distinct points of T, then $xy = \{x, y\}$ is a subset of T. Since $V(T)$ is the smallest linear variety which contains T, we have $V(T) = T$. Note that the only set which generates $V(T)$ is T itself.

Example 7 If R^3 is the set of points associated with Euclidean solid geometry and P, Q, and R are distinct noncollinear points of R^3, then $V(\{P, Q, R\})$ is PQR, the plane determined by P, Q, and R. Any subset of PQR which contains any three noncollinear points generates PQR. No subset of PQR containing fewer than three noncollinear points generates PQR.

Proposition 5 *If x and y are distinct points of a linear space S, then $V(\{x, y\}) = xy$.*

Proof xy is a linear variety which contains x and y, yet every linear variety which contains x and y must contain xy as well.

We are now in a position to define the important notions of *independence* and *dimension*.

Definition 5 *Let S be a linear space. A linear variety V in S is said to have* **dimension n**, *or to be* **n-dimensional**, *if V is the linear closure of some subset T of S which contains n + 1 points of S, but is not the linear closure of any subset of S which contains less than n + 1 points; that is, V is generated by n + 1 points, but not by less than n + 1 points.*

A set T of n + 1 points of S is said to be **independent** *if T is not contained in any variety of dimension less than n.*

We will use dim V to denote the dimension of a linear variety V.

The proofs of the following propositions are left as exercises.

Proposition 6 *The dimension of any one point set is 0; the dimension of any line is 1.*

Proposition 7 *Any subset of an independent set is independent.*

While we do have some of the properties of independence and dimension in linear spaces that we might expect to have, based on our previous experience with vector spaces or Euclidean geometry, we also fail to have certain properties that might reasonably be expected to apply to these notions as we see from the next example.

Example 8 Suppose $\{x, y, z\}$ is an independent subset of a linear space S, and u, v, and w are three noncollinear points of $V(\{x, y, z\})$. We might expect to have

$$V(\{x, y, z\}) = V(\{u, v, w\}).$$

Such would be the case with Euclidean geometry, for example, since a plane is determined by *any* three of its noncollinear points. We now give an example of a linear space for which our expectation that

$$V(\{x, y, z\}) = V(\{u, v, w\})$$

is not fulfilled.

Let S be the triangle ABC as shown in Fig. 1.5. Each element of the triangle will be a point. If x and y are distinct points of S, then we let $xy = \{x, y\}$ if x and y are on different sides of ABC, and let xy be the side of ABC containing x and y if x and y are on the same side. With lines thus defined, S is a linear space. Now

$$V(\{A, B, C\}) = S.$$

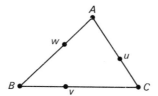

Fig. 1.5

Let u, v, and w be three points of S chosen from distinct sides of ABC, but let none of these points be equal to A, B, or C. Then

$$V(\{u, v, w\}) = \{u, v, w\},$$

which is not the same as $V(\{A, B, C\})$ even though u, v, and w are non-collinear points of $V(\{A, B, C\})$.

An even more striking example of a property that conflicts with our intuitive notion of what a geometry should be like is discussed in Exercise 3 below.

In linear spaces we have some, but by no means all, of the properties that we might feel reasonable geometries should have. The reader might try to find out for himself the additional assumptions about a linear space that would ensure having:

(1)   A variety V of dimension n is generated by any independent subset of V which contains $n + 1$ points.

The reader might also investigate what effect taking (1) as an axiom would have on the structure of a linear space. It is of such stuff that the study of geometry is made.

EXERCISES

1. Prove Proposition 6.

2. Prove Proposition 7.

3. Let S be the subset of the usual coordinate plane R^2 described as follows: S will consist of the triangle with vertices $(2, 0)$, $(-2, 0)$, and $(0, 2)$ together with the portions of the lines with equation $y = 1/n$, n any positive integer, which lie on or inside the triangle (see Fig. 1.6). Any element of S will be a point. Given two points P and Q of S, we define the line PQ as follows: If P and Q are both contained in the same side of the triangle or some one of the cross lines, we let the side of the triangle or the cross line be PQ. Otherwise, $PQ = \{P, Q\}$. Prove that S with the structure so defined is a linear space. Prove that $\dim S = 3$. Prove that given any positive integer m, S contains a linear variety of dimension m.

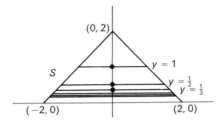

4. Below are a number of statements about linear spaces. If a statement is true, supply a proof. If a statement is false, supply an example of a linear space in which the statement fails to hold.

a) The dimension of any linear space is at least 2.

b) If V and W are linear varieties, then $\dim V(V \cup W) \leq \dim V + \dim W + 1$.

c) If S is a linear space and $\dim S = 3$, then the intersection of any two 2-dimensional linear varieties of S must either be empty or have dimension 1. (Think about planes in Euclidean solid geometry. Any two planes which intersect, intersect in a line.)

d) If S is a linear space, S' a subset of S which contains at least two points, and if S' has the linear structure induced by S (cf. Example 5), then if T is a subset of S', the linear closure of T relative to S' is the intersection of $V(T)$(in S) with S'; that is $V_{S'}(T) = V(T) \cap S'$.

5. Prove the assertions made in Examples 2 and 6.

6. Construct all linear spaces which have two points, which have three points, and which have four points.

7. Answer the question posed in Example 5; that is, in defining the linear structure induced on S' by S, why would it not suffice to let the lines of S' be the intersections of lines in S with S'? If we did define a geometric structure on S', letting the lines of S' be such intersections, what properties of a linear space would not be satisfied?

1.3 ISOMORPHISMS. COLLINEATIONS

Whenever a mathematician deals with structured sets, he is interested in functions between the structured sets which respect the structure in question. For example, if G, $\#$ and H, $*$ are groups, then the structure-respecting functions from G into H are the *homomorphisms* from G into H, that is, those functions f such that

$$f(g \# g') = f(g) * f(g') \qquad \text{for each } g \text{ and } g' \text{ in } G.$$

(Observe that a homomorphism carries the "product" of g with g' into the product of $f(g)$ with $f(g')$, which is why we say that f is structure-respecting.)

The mathematician is also interested in functions between structured sets which not only respect structure, but preserve it; such functions are called *isomorphisms*. An isomorphism between a group G, $\#$ and a group H, $*$ is a function f from G into H which is one-one and a homomorphism. It can be shown that requiring f to be one-one, onto, and a homomorphism automatically makes f^{-1} a homomorphism from H onto G. Thus, an isomorphism from G onto H is a one-one function which is structure-respecting and which has a structure-respecting inverse. Two groups which are isomorphic are identical as groups and differ at most in the way each is labeled. For example, if f is an isomorphism from G, $\#$ onto H, $*$, then relabeling $\#$ by $*$ and each g in G by $f(g)$, we make the group G, $\#$ indistinguishable from H, $*$. We would not have changed any group property of G, $\#$ in the relabeling; we would merely have made G, $\#$ look like H, $*$.

If we restrict our attention to those linear spaces S such that each element of a linear space S is a point, the geometric structure of S is determined entirely by which subsets of S are lines. A structure-respecting function f from one such linear space S into another such linear space S' would therefore be one which takes lines to lines; that is, for each line L of S,

$$f(L) = \{f(s) \mid s \text{ is a point of } L\}$$

is a line in S'.

Note: We will henceforth assume that the points of any linear space S are precisely the one-element subsets of S. All important spaces are of this type, or can be expressed in this form.

Definition 6 *A function f from a linear space S into a linear space S', having the property that $f(L)$ is a line of S' whenever L is a line of S, is said to be a* **linear function***.*

Linear functions have the following important properties.

Proposition 8 *Suppose f is a linear function from the linear space S into the linear space S'. Then:*

a) *If V is a linear variety in S, then $f(V)$ is a linear variety in S'.*
b) *If V' is a linear variety in S', then $f^{-1}(V')$ is a linear variety in S.*

Proof a) Let y and y' be any two distinct points of $f(V)$. Then we have distinct points x and x' of V with $f(x) = y$ and $f(x') = y'$. Since V is a linear variety, xx' is a subset of V; hence $f(xx')$ is a subset of $f(V)$ which contains both y and y'. But since f is linear, $f(xx')$ is a line in S'; hence it

---

* Linear functions are to geometry what homomorphisms are to the study of groups; indeed, some texts use the term *homomorphism* in place of *linear function*. Linear functions are not as important in the study of geometry, however, as homomorphisms are in group theory.

must be the line yy'. Therefore yy' is a line in $f(V)$ and $f(V)$ is a linear variety.

We leave the proof of (b) as an exercise.

The proof of (a) of Proposition 8 also shows

Proposition 9 *If f is a linear function from a linear space S into a linear space S' and $f(x) \neq f(x')$, then*

$$f(xx') = f(x)f(x').$$

Definition 7 *A one-one and onto linear function from a linear space S into a linear space S' is said to be an **isomorphism** from S onto S'. If there is an isomorphism from S onto S', then we say that S and S' are isomorphic.*

*An isomorphism of a linear space S onto itself is called a **collineation**.* (Such a function is called an *automorphism* in some texts.)

If two linear spaces are isomorphic, then from a geometrical point of view they are equivalent. By properly relabeling one of the spaces, the spaces can be made indistinguishable.

We will soon see that if f is an isomorphism from S onto S', then f^{-1} is an isomorphism from S' onto S. This result will follow as a corollary of the following proposition.

Proposition 10 *Suppose that f is an isomorphism from a linear space S onto a linear space S'. Then,*

a) *If V is a linear variety in S, $\dim f(V) = \dim V$.*
b) *If V' is a linear variety in S', then $\dim f^{-1}(V') = \dim V'$.*

Proof a) Suppose that $\dim V = n$ and T is a set of $n + 1$ points which generates V. Then $f(T)$ is a set of $n + 1$ points of S' since f is one-one, and $f(T)$ generates $f(V)$. For if $f(T)$ did not generate $f(V)$, then $V(f(T))$ would be a linear variety of S' contained in but not equal to $f(V)$. Therefore $f^{-1}(V(f(T)))$ would be a linear variety in S contained in but not equal to V (Fig. 1.7). But, in that case, since T is a subset of $f^{-1}(V(f(T)))$, T could not generate V. Since a set of $n + 1$ points generates $f(V)$, we have

$$\dim f(V) \leq n = \dim V.$$

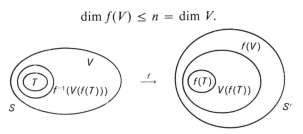

Fig. 1.7

Suppose now that some set Y generates $f(V)$. Then an argument similar to that used to show that $f(T)$ generates $f(V)$ proves that $f^{-1}(Y)$ generates $f^{-1}(f(V)) = V$. Consequently, since dim $V = n$, $f^{-1}(Y)$ (and hence, Y itself, since f is one-one) contains at least $n + 1$ points. Therefore dim $f(V) \geq n =$ dim V. Consequently, dim $V = n =$ dim $f(V)$.

The proof of (b) is left as an exercise.

Corollary *If f is an isomorphism from the linear space S onto the linear space S', then f^{-1} is also an isomorphism from S' onto S.*

Proof f^{-1} is one-one and onto, since f is one-one and onto. It must be shown that f^{-1} is linear. If L' is a line in S', then $f^{-1}(L')$ is a linear variety of dimension 1 in S. A linear variety of dimension 1, however, is a line; hence $f^{-1}(L')$ is a line in S if L' is a line in S'. Therefore f^{-1} is linear.

Proposition 11

a) *Any linear space is isomorphic to itself.*

b) *If a linear space S is isomorphic to a linear space S', then S' is isomorphic to S.*

c) *If S, S', and S'' are linear spaces with S isomorphic to S' and S' isomorphic to S'', then S is isomorphic to S''.*

This proposition says that "is isomorphic to" determines an equivalence relation on the class of all linear spaces; an *isomorphism class* consists of all spaces isomorphic to one another, that is, of all spaces equivalent from the viewpoint of their linear geometric structure.

Proof

a) The identity function on a linear space S is an isomorphism from S onto itself.

b) If f is an isomorphism from S onto S', then f^{-1} is an isomorphism from S' onto S.

c) If f is an isomorphism from S onto S' and g is an isomorphism from S' onto S'', then $g \circ f$, the composition of g with f, is an isomorphism from S onto S''. For, the composition of two one-one and onto functions is one-one and onto. Moreover, if L is a line in S, then $f(L)$ is a line in S'. But then

$$g\big(f(L)\big) = (g \circ f)(L)$$

is a line in S''; hence $g \circ f$ is linear.

The next proposition hints that geometric axioms may also have algebraic implications.

Proposition 12 *If S is a linear space, then the set of collineations of S is a group with composition of functions as the group operation.*

Proof Denote the set of collineations of S by $G(S)$. The composition of any two collineations of S is again a collineation of S (by the proof of (c) in Proposition 11 with $S = S' = S''$); hence $G(S)$ is closed with respect to $\circ$. The identity collineation of S serves as the group identity of $G(S)$, $\circ$. Composition of functions is always associative. Finally, if f is a collineation of S, then f^{-1} is also a collineation of S and is the inverse with respect to $\circ$ of f. Therefore $G(S)$, $\circ$ satisfies all of the axioms for a group.

Definition 8 *The set of collineations of a linear space S with function composition as the operation is called the **collineation group** of S. We continue to denote this group by $G(S)$, $\circ$, or merely $G(S)$.*

The algebraic properties of $G(S)$, $\circ$ evidently depend heavily on the geometric properties of S. We might reasonably expect, then, that certain geometric statements about S can be rephrased as algebraic statements about $G(S)$. In point of fact, some linear spaces can be characterized entirely by the algebraic properties of their collineation groups. We conclude this chapter by considering one aspect of the relationship of a linear space with its collineation group.

Definition 9 *Let S be a linear space. Then $G(S)$, $\circ$ is said to be **transitive with respect to a family** $\mathcal{A}$ **of subsets of S** if, given any two members T and T' of $\mathcal{A}$, there is a collineation f of S such that $f(T) = T'$.*

*If $G(S)$ is transitive with respect to the family of one-point subsets of S, we say that $G(S)$ is **transitive**, and that S is **homogeneous**.*

Example 9 Euclidean plane geometry is homogeneous. It may be shown—although we will not show it—that any collineation f of the coordinate plane R^2 has the form

$$f(x, y) = (ax + by + c, a'x + b'y + c'),$$

where a, b, c, a', b', and c' are real numbers and $ab' - a'b \neq 0$. If we want a collineation which takes the point (x_1, y_1) into the point (x_2, y_2), we need to solve the equations

$$x_2 = ax_1 + by_1 + c, \quad \text{and} \quad y_2 = a'x_1 + b'y_1 + c'$$

simultaneously subject to the condition $ab' - a'b \neq 0$. There are, in fact, an infinite number of solutions to this system of equations; hence the usual coordinate plane with its geometry is homogeneous.

Example 10 Consider the linear space S described in Example 8. There is no collineation of S which takes a point in the interior of some side of ABC, say the point u (Fig. 1.5), into a vertex, say C. For any collineation of S must take lines into lines, and, hence, will take a side of ABC into a side of ABC.

This fact, in turn, implies that any collineation of S takes the vertices of ABC into the vertices; therefore S is not homogeneous.

If there is a collineation f of a linear space S which takes a subset T of S onto a subset T' of S, then, from the viewpoint of the linear structure of S, T and T' can be considered to be equivalent (just as two isomorphic linear spaces can be considered to be equivalent). In a homogeneous space, any two points are interchangeable. In Example 8 the vertices of ABC are not geometrically equivalent to points which are not vertices.

If a linear space S is transitive with respect to its set of lines, then, given any two lines L and L' of S, there is a collineation f of S such that $f(L) = L'$. This tells us that any two lines of S are linearly equivalent; for example, they contain the same number of points and meet the same number of lines. We can see that the linear space of Example 8 is not transitive with respect to its family of lines since no two-point line could possibly be carried by a collineation of S onto a side of ABC. The usual coordinate plane with its Euclidean geometry is transitive with respect to its set of lines; the proof of this fact is left as an exercise. We intuitively feel that any two lines of analytic geometry are essentially the same from a geometric point of view, even though they may have certain properties not intrinsically connected with the linear space structure (such as their slopes) which are different.

This chapter has introduced a basic geometry and some of the general questions associated with the study of geometry. In the next chapter, we will restrict the geometry under consideration by certain assumptions in addition to those for a linear space, and we will investigate the consequences of the added restrictions.

EXERCISES

1. In Example 8, prove that u cannot be mapped by a collineation of S onto C. Prove or disprove the following: Any collineation of S in Example 8 is uniquely determined by what it does to A, B, and C.

2. Prove (b) of Proposition 8.

3. Prove (b) of Proposition 10.

4. Prove that, given any two lines in plane analytic geometry, there exists a collineation of the coordinate plane which maps one line onto the other. You may use the information given in Example 9 if you wish.

5. Decide which of the following statements are true and which are false. If a statement is true, prove it. If a statement is false, supply a counterexample of a linear space to which the statement does not apply.

   a) If f is a linear function from a linear space S into a linear space S' and V is a linear variety in S, then dim $V \geq$ dim $f(V)$.

   b) If f is a linear function from a linear space S into a linear space S' and V' is a linear variety in S', then dim $V' \leq$ dim $f^{-1}(V')$.

c) If f is a linear function from a linear space S into a linear space S' and T is a subset of S, then $f(V(T)) = V(f(T))$.

d) The set of collineations of a linear space S which leave the point s of S fixed form a subgroup of collineations of S. (A collineation f leaves s fixed if $f(s) = s$.)

e) There are precisely two collineations of any linear space containing just two points.

6. Describe the collineation group of the space presented in Example 2. Is this space homogeneous? Is it transitive with respect to its family of lines?

7. Find all collineations of a linear space which contains three points and three lines. Construct a multiplication table for this group of collineations.

2 Projective and Affine Planes

2.1 PROJECTIVE PLANES

When Euclid described the geometry which is now called by his name, he believed that he was dealing with the geometry of the real world. The axioms for Euclidean geometry were drawn from direct observation of the physical universe, and the terms of the geometry referred to real objects. While mathematicians today admit that the real world can inspire the definition of abstract mathematical systems, and mathematical systems can be used to draw conclusions about the world, mathematical systems today are recognized to exist in their own right and need not describe a real phenomenon to justify their existence. We can draw logical conclusions in a mathematical system without ever concerning ourselves about the existence of a real model for the system. However, since the existence of a model for a mathematical system at least assures us that the axioms of the system are not self-contradictory, it is reassuring to have at least one model.

The Greeks looked at the physical universe and included what they thought were its geometric properties in a body of knowledge that came to be known as Euclidean geometry. We shall abstract our definition of the projective plane from real observations, but we will find that once the abstraction has been carried out, we will be able to prove a good deal about projective planes which is jarring to the intuition. We will abstract from reality and find that when we extract the logical consequences from our abstraction, we often seem to be dealing with something which has no basis in reality. Whether or not this disturbs us will depend on how willing we are to play

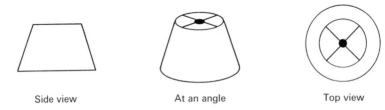

| Side view | At an angle | Top view |

Fig. 2.1

the mathematical game for its own sake. Once a mathematician defines the system he will work in, he must abide by the logical consequences of that system, no matter how distasteful they may be to him, unless he wants to start all over again by redefining the system.

It is well known that when we look down straight railroad tracks, the parallel rails seem to meet in the distance. This phenomenon of seeing things in perspective was not incorporated into Western art until the Renaissance. Before that time artists' depictions were flat and strictly two-dimensional. They lacked the depth, or third dimension, that we perceive in nature. With the incorporation of perspective into art came an interest in studying the principles involved in the phenomenon of perspective. For example, consider the lampshade of Fig. 2.1. The bottom edge of the lampshade may be a perfect circle, but how it looks to us will depend on the angle from which we view it. If we view it from the side, the bottom edge may appear to be a straight line or an ellipse. In fact, it will appear in its true circular shape only if we look at it from directly above or directly below. The point is that the appearance of a geometrical object is determined by our viewpoint. The appearance of an object at some given angle is a question of great importance to the artist; hence, it is not surprising that extensive investigation on this matter was conducted even while modern mathematics was still in its infancy, or that much of this investigation was carried out by artists.

We now try to express axiomatically at least some of the properties of our perspective vision. We restrict ourselves for simplicity to two dimensions, that is, to our view of things on a flat surface. We shall begin with a linear space and two additional assumptions.

Definition 1 *A linear space S is said to be a **projective plane** if*:*

a) *Any two lines in S contain a point in common.*

b) *There are four points, no three of which are collinear.*

Property (a) of a projective plane embodies the perspective phenomenon of all lines appearing to meet. Property (b) is included to avoid certain

---

* We continue the assumption throughout that the points of S are the same as the one element subsets of S.

trivial cases. To summarize, a projective plane is a set S whose elements are the points of the plane, and a collection of subsets of S, called lines, such that any two distinct points of S are contained in a unique line, and any two distinct lines contain a common point. Moreover, any line contains at least two points, and S contains some subset which contains four points, no three of which are collinear.

Notation and conventions Unless explicitly specified to be otherwise, any proposition in this chapter will refer to a projective plane π. Points of π will be denoted by the capital roman letters P, Q, R, and U (with primes and/or subscripts if needed), and lines will be denoted by the capital roman letters H, L, and M (with primes and/or subscripts if needed).

We now begin to prove some of the basic properties of projective planes.

Proposition 1 *Any line contains at least three points.*

Proof Let L be any line and P and Q be distinct points of L. Since π contains four points, no three of which are collinear, it follows that there must be two points P' and Q' of π which are not collinear with either P or Q (Exercise 7). Then $P'Q'$ meets L in some point other than P or Q. Hence L contains P, Q, and at least one other point (Fig. 2.2).

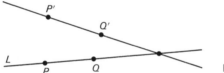

Fig. 2.2

The following example shows that a projective plane may have lines which contain only three points.

Example 1 Set $\pi = \{1, 2, 3, 4, 5, 6, 7\}$. We let the lines of π be

$$\{1, 2, 3\}, \quad \{3, 4, 5\}, \quad \{1, 5, 6\}, \quad \{1, 4, 7\}, \quad \{3, 6, 7\}, \quad \{2, 5, 7\}, \quad \text{and} \quad \{2, 4, 6\}.$$

A graphic representation of the lines and points and their relationships is given in Fig. 2.3. The reader should verify that π satisfies the axioms for a projective plane.

The reader might note several interesting facts about the projective plane of this example. In the first place, all of the lines have precisely the same number of points, three. Second, there is no linear variety of dimension greater than 2, and π is the only linear variety of dimension 2. There are other interesting properties of this projective plane, but some of these will not be evident until we have studied projective planes in greater depth.

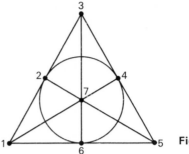

Fig. 2.3

Before continuing with proofs of basic facts about projective planes, we give another example of a projective plane, or, rather, a method of obtaining many projective planes.

Example 2 Let V be a 3-dimensional vector space over a field F. (Recall that in such an instance V can be considered to be the set F^3 of all ordered triples of elements of F with coordinate-wise vector addition and scalar multiplication.) Let π be the set of 1-dimensional subspaces of V; π is the set of points for our projective plane. Given two distinct points P and Q of π (that is, two 1-dimensional subspaces of V), we define the line PQ to consist of all points of π contained in the 2-dimensional subspace of V which contains P and Q. If $\{a\}$ and $\{b\}$ are bases for P and Q, respectively, then $\{a, b\}$ is a basis for the 2-dimensional subspace which contains P and Q.

We now prove that π with the linear structure described above is a projective plane. It is fairly evident that any line contains at least two points (that is, any 2-dimensional subspace contains at least two 1-dimensional subspaces), and any two distinct points are contained in a unique line. We must still show that any two lines contain a common point and that π contains some four points, no three of which are collinear.

A classical theorem of linear algebra states that given subspaces W and W' of a vector space V, then

(1) $\dim W + \dim W' = \dim (W \cap W') + \dim (W + W')$.

If W and W' are 2-dimensional and distinct and if V is 3-dimensional, then (1) gives

(2) $2 + 2 = \dim (W \cap W') + 3;$

consequently, $\dim (W \cap W') = 1$. It follows, then, in the case under consideration, that the intersection of any two lines is a point.

The existence of four points, no three of which are collinear, is equivalent to having four 1-dimensional subspaces of V, no three of which are contained in the same 2-dimensional subspace of V, which, in turn, is equivalent to having four vectors in V, any three of which form a basis for V. We leave the proofs of the stated equivalences to the reader. We also leave as an exercise a proof that $(1, 0, 0)$, $(0, 1, 0)$, $(0, 0, 1)$, and $(1, 1, 1)$ are in fact four vectors in V (regardless of what field F is), any three of which form a basis for V.

Definition 2 *We will call the projective plane associated with the vector space V, as described in Example 2, the* **projective plane associated with** V. *We will denote such a projective plane by $\pi(V)$.*

The projective planes associated with vector spaces are of such great importance that we will return to them several times during the course of this text. For the moment, we merely give a specific example of such a projective plane.

Example 3 Let V be the 3-dimensional vector space over Z_2, the field of integers modulo 2 (V can be associated with Z_2^3). Then V contains eight elements:

$$(0, 0, 0), \ (0, 0, 1), \ (0, 1, 0), \ (1, 0, 0), \ (0, 1, 1), \ (1, 0, 1), \ (1, 1, 0), \ (1, 1, 1).$$

There are seven 1-dimensional subspaces of V; a 1-dimensional subspace in this instance consists of $(0, 0, 0)$ and exactly one nonzero element. (This follows from the fact that the sum of any nonzero element v with itself is the zero vector, and the only nonzero scalar multiple of v is v itself.) Consequently, $\pi(V)$ contain seven points. There are also seven 2-dimensional subspaces of V; hence there are seven lines in $\pi(V)$. Explicitly, the 2-dimensional subspaces of V are

$\{(0, 0, 0), (1, 0, 0), (0, 1, 0), (1, 1, 0)\}, \quad \{(0, 0, 0), (0, 0, 1), (1, 0, 0), (1\ 0, 1,)\},$
$\{(0, 0, 0), (0, 0, 1), (1, 1, 1), (1, 1, 0)\}, \quad \{(0, 0, 0), (0, 1, 0), (0, 1, 1), (0, 0, 1)\},$
$\{(0, 0, 0), (1, 1, 0), (1, 0, 1), (0, 1, 1)\}, \quad \{(0, 0, 0), (0, 1, 0), (1, 1, 1), (0, 1, 0)\},$
$\{(0, 0, 0), (1, 1, 1), (1, 0, 0), (0, 1, 1)\}.$

We have here, as in Example 1, a projective plane with seven points and seven lines; the reader should try to determine if these two projective planes are isomorphic.

It can be proved fairly readily that a projective plane of the type $\pi(V)$ does not contain any linear variety other than itself of dimension greater than or equal to 2. We now prove that this property holds for all projective planes; hence a projective plane is generated by any three of its noncollinear points.

Proposition 2 *If P, Q, and R are three noncollinear points of π, then $V(\{P, Q, R\}) = \pi$.*

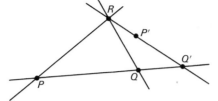

Fig. 2.4

Proof The linear variety $V(\{P, Q, R\})$ contains the lines PQ, PR, and QR. Suppose P' is any point of π not on one of these lines. We will show that P' is contained in $V(\{P, Q, R\})$. Consider the line $P'R$. The lines $P'R$ and PQ share a common point Q' (Fig. 2.4). Now $Q' \neq R$ since R is not a point of PQ, but Q' is. Therefore $Q'R$ is a line which contains R, Q', and P', and $P'R = Q'R$. Now Q' and R are distinct points of $V(\{P, Q, R\})$, a linear variety; hence $Q'R$ is a subset of $V(\{P, Q, R\})$. But this implies that P' is a point of $V(\{P, Q, R\})$. It follows then that any point of π is a point of $V(\{P, Q, R\})$; consequently, $\pi = V(\{P, Q, R\})$.

EXERCISES

1. Draw a graphical representation of the projective plane of Example 3. Find an isomorphism from that projective plane onto the projective plane described in Example 1.

2. Prove the equivalences given for the existence of four points, no three of which are collinear, given in Example 2. In that same example, prove that $(1, 0, 0)$, $(0, 1, 0)$, $(0, 0, 1)$, and $(1, 1, 1)$ have the property that any three of these vectors form a basis for V.

3. What are the points of the projective plane $\pi(Z_3^3)$? How many lines do you think this plane has?

4. Prove that any projective plane has at least seven points.

5. Prove that the linear space described in Example 4 of Chapter 1 is a projective plane. Can you associate this projective plane with a projective plane of the type described in Example 2 of this chapter?

6. Which of the following linear spaces are projective planes? If a linear space fails to be a projective plane, indicate precisely which properties of a projective plane the linear space does not have.
   a) π is the vertices of a triangle and the lines are pairs of vertices.
   b) π is the vertices of a square and the lines are pairs of vertices.
   c) π is the usual Euclidean plane with its ordinary geometry.

7. Prove that given any two points P and Q of a projective plane π, there exist points R and S in π such that no three of the points P, Q, R, and S are collinear.

2.2 BASIC PROPERTIES OF A PROJECTIVE PLANE

Proposition 3 *There are at least four lines, no three of which contain a common point.*

Proof By (b) of Definition 1 we have four points P, Q, R, and U, no three of which are collinear. We now prove that no three of the lines PQ, QR, RU, and PU share a common point. Suppose that some three of these lines did in fact share a common point; we may assume the three lines are PQ, QR, and RU. Since PQ and QR already share Q in common, the common point must be Q. But then Q is a point of RU; hence Q, R, and U are collinear, contradicting the assumption that no three of the points P, Q, R, and U are collinear. Therefore no three of the lines PQ, QR, RU, and PU can share a point in common.

For the sake of ready reference we restate the axioms for a projective plane. The statements are not precisely the same as the earlier statements, but they are equivalent to them.

Axiom 2.1 *Any two distinct points are contained in a unique line.*

Axiom 2.2 *Any two distinct lines contain a unique point.*

Axiom 2.3 *There are four points, no three of which are contained in the same line.*

Axiom 2.4 *Each line contains at least two points.*

The next proposition shows that Axiom 2.4 is implied by Axioms 2.1, 2.2, and 2.3, and hence is superfluous.

Proposition 4 *If S is a set of points with a family of lines which satisfies Axioms 2.1, 2.2, and 2.3, then each line contains at least two points.*

Proof Suppose L is a line in S which contains but a single point P. Since every line of S meets L in some point, it follows that every line of S contains P. But Proposition 3, which was proved without the use of Axiom 2.4, tells us that S contains four lines, no three of which contain a common point. Therefore S cannot contain a line consisting of exactly one point. No line L can be the empty set since then no line could intersect L in a point. Therefore, any line must contain at least two points.

Corollary *A projective plane is characterized by Axioms 2.1, 2.2, and 2.3.*

The following proposition shows that Axiom 2.3 and Proposition 3 are equivalent if Axioms 2.1 and 2.2 are assumed. That is, if we take Axioms 2.1 and 2.2, and Proposition 3, as our axioms, then Axiom 2.3 can be proved as a theorem. This proposition illustrates that a mathematical system may have several different (although equivalent) axiomatic foundations.

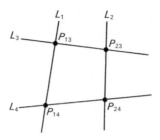

Fig. 2.5

Proposition 5 *If we assume Axioms 2.1 and 2.2, then Proposition 3 is true if and only if Axiom 2.3 is true.*

Proof We have already shown that Axioms 2.1, 2.2, and 2.3 imply Proposition 3. Suppose now that Proposition 3 is true. Then there are lines L_1, L_2, L_3, and L_4, no three of which contain some common point. We denote the point common to L_i and L_j, $i \neq j$, by P_{ij}. Consider the four points P_{13}, P_{23}, P_{24}, and P_{14} (Fig. 2.5). We will show that no three of these points are collinear. Suppose that three of these points are collinear; by proper labeling, we can suppose the three collinear points are P_{13}, P_{23}, and P_{24}. All of these points must be distinct; for if two of them were equal, say $P_{13} = P_{23}$, then P_{13} would be a point common to L_1, L_2, and L_3, contradicting the assumption that these lines share no point in common. But if these points are distinct, then we have $P_{13}P_{23} = L_3 = P_{23}P_{24} = L_2$; hence $L_2 = L_3$, again a contradiction. Therefore the four points P_{13}, P_{23}, P_{24}, and P_{13} must be such that no three of these points are collinear.

Corollary *A set S of points, which has a collection of subsets called lines, is a projective plane if and only if the points and lines satisfy these properties: Axioms 2.1 and 2.2 as stated before, and the following axiom.*

Axiom 2.3′ *There are four lines, no three of which contain a common point.*

Although what we have done thus far has a certain interest because it shows that a mathematical system may have several axiomatizations, its implications are actually far more significant and far-reaching. Suppose we take Axioms 2.1, 2.2, and 2.3 and interchange *point* with *line* and *contains* with *is contained in*. Interchanging the terms in this manner we obtain the following statements

(3)   Any two distinct lines contain a unique point.

(4)   Any two distinct points are contained in a unique line.

(5)   There are four lines, no three of which contain a common point.

Observe that (3) is nothing but Axiom 2.2, (4) is Axiom 2.1, and (5) is Proposition 3, which, in the context of Axioms 2.1 and 2.2, is equivalent to Axiom 2.3. In other words, interchanging the terms in Axioms 2.1, 2.2, and 2.3 as indicated, we again have axioms for a projective plane. We now introduce a term which will enable us to express concisely the effect of what we have observed.

Definition 3 *If $\mathcal{P}$ is any statement about a projective plane, then the statement formed from $\mathcal{P}$ by interchanging the terms* **line** *and* **point**, *and* **contains** *and* **is contained in***, is called the* **dual statement of** $\mathcal{P}$.

We thus see that Axioms 2.1 and 2.2 are dual statements of one another, while Proposition 3 and Axiom 2.3 are dual statements. We have observed that if we take the dual statements of the axioms for a projective plane, we again obtain an equivalent set of axioms for a projective plane. The importance of the equivalence of the axioms and their dual statements rests in the fact that this equivalence means that if any statement is true about a projective plane, then its dual statement is also true. Thus, whenever we prove any statement about a projective plane, we generally obtain another statement free of charge, namely, the dual of the statement we prove. (We say generally not because the dual statement about the projective plane is ever false, but because sometimes the dual turns out to be essentially the same statement; hence the dual does not always give us new information.) This interchangeability of the terms described is known as the *Principle of Duality*. Using the Principle of Duality alone we can make the following statements.

Proposition 6

a) *If there are n points in a projective plane, then there are n lines.* (That is, if the statement, "There are *n* points," is true, then its dual, "There are *n* lines," is also true.)

b) *If there are n lines, then there are n points.*

c) *If each point is contained in m lines, then each line contains m points.*

d) *If each line contains m points, then each point is contained in m lines.*

Note that (a) and (b) are dual statements, as are (c) and (d).

Proposition 7 *Each point is contained in at least three lines. Dually, each line contains at least three points.*

Proof Let P be any point. Because there are four points, no three of which are collinear, we can find Q, R, and U such that P, Q, R, and U are four points, no three of which are collinear (Exercise 1, Section 2.1). Then P is contained in the three lines PQ, PR, and PU, and all of these lines are distinct.

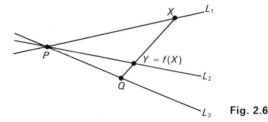

Fig. 2.6

Proposition 8 *Any two lines contain the same number of points.*

Proof Let L_1 and L_2 be any two lines. Then L_1 and L_2 contain a common point P. By Proposition 7 there is at least one other line L_3 which contains P, and there are at least three points in L_3. We can therefore choose Q a point of L_3 different from P; clearly Q is not a point of either L_1 or L_2. Let X be any point which is contained in L_1. Then QX and L_2 contain a common point Y; we set

$$f(X) = Y$$

(Fig. 2.6). In this manner we define a function f from L_1 into L_2. We now prove that f is one-one and onto.

Suppose $f(X) = f(X')$. Then $Qf(X) = Qf(X')$. Now X is the point common to L_1 and $Qf(X)$, and X' is the point common to L_1 and $Qf(X')$. Consequently, $X = X'$; therefore f is one-one. Suppose Y is any point of L_2. Then QY and L_1 share a common point X. Consequently, $Y = f(X)$; hence f is onto.

Since we have been able to produce a one-one function from L_1 onto L_2, L_1 and L_2 have the same number of points.

As a corollary of Proposition 8 we have the following dual statement.

Corollary *Any two points are contained in exactly the same number of lines.*

If each line contains an infinite number of points, then obviously, the projective plane contains an infinite number of points. We may ask, however, how many points there are if each line contains precisely n points. (A projective plane with a finite number of points may jar the intuition, but nothing in the axioms for a projective plane forbids it. Indeed, we have already seen projective planes with finitely many points.)

Proposition 9 *If each line contains n points, then there are exactly $n^2 - n + 1$ points.*

Proof Let L be any line and P be any point which is not contained in L. There are n distinct lines of the type PX, where X is some point contained in L; that is, there is one such line for each point of L. Each of these n lines contains n points and P is the only point that they all share in common. Any point is contained in at least one of these lines; since if Y is any point, then PY and L contain some point X. Hence Y is contained in PX. If all lines of the form PX had no points in common, we would have precisely n^2 points. But all of these lines contain P; hence to avoid counting P n times, we must take $n - 1$ (one for each of the lines except one) from n^2. This procedure leaves us with a total of $n^2 - (n - 1) = n^2 - n + 1$ points.

Corollary (*the dual of Proposition 9*) *If each point is contained in n lines, then there are exactly $n^2 - n + 1$ lines.*

Proposition 10 *If each line contains n points, then there are $n^2 - n + 1$ points and $n^2 - n + 1$ lines.*

This proposition follows at once from Proposition 6.

Definition 4 *A projective plane, one of whose lines (and hence any one of whose lines) contains $m + 1$ points, is said to be of* **order** *m.*

The following is an immediate corollary of Definition 4 and Proposition 10.

Corollary *A projective plane of order m contains $m^2 + m + 1$ points and $m^2 + m + 1$ lines.*

We conclude, then, there can be no projective plane with 5 or 6 points, since neither 5 nor 6 is of the form $n^2 + n + 1$ for any positive integer n. If $n = 2$, then $n^2 + n + 1 = 7$; hence there might be a projective plane of order 2 with 7 points and 7 lines. Such a plane was given in Example 1. We might ask whether, given some positive integer n, there is a projective plane of order n. The answer to this question is still one of the great unsolved problems of mathematics. Another unsolved problem is the following: If there is at least one projective plane of order n, how many nonisomorphic projective planes of that order are there? Only partial answers to both of these questions have been found, and complete answers seem a long way off.

EXERCISES

1. It is possible to have a projective plane of 13 points and 13 lines, that is, a projective plane of order 3. Construct such a projective plane using as points the integers 1 through 13. Construct such a plane using the vector space Z_3^3.

2. Which of the following statements are true and which are false? If a statement is true, supply a proof; if false, give a counterexample.

   a) Any projective plane containing a finite number of points contains an odd number of points.

b) Any two points of a projective plane are contained in exactly the same number of lines.

c) There is a projective plane containing 41 points.

d) If the order of a projective plane is m, then each point is contained in exactly $m + 1$ lines.

3. Explain carefully why the Principle of Duality is true for projective planes. That is, why is a statement about a projective plane true if and only if its dual statement is true? Consider the following: Suppose $\mathscr{A}$ and $\mathscr{A}'$ are systems of axioms, both involving the terms T and T'. Suppose that when T is interchanged with T' in $\mathscr{A}$, we get $\mathscr{A}'$. If $\mathscr{P}$ is a proposition that can be proved from $\mathscr{A}$, and T and T' are interchanged in $\mathscr{P}$ to give a new proposition $\mathscr{P}'$, is $\mathscr{P}'$ provable from $\mathscr{A}'$? How? Relate these observations to what was done in this section concerning duality in the projective plane.

4. Is the Principle of Duality true relative to ordinary Euclidean geometry? Consider: Given a point P which is not contained on a line, there is precisely one line which contains P and does not share a common point with L. What is the dual of this statement? Is it true in the context of Euclidean geometry?

5. Try to formulate axioms for a projective solid geometry. Our geometric objects would be points, lines, and planes. We would want, among other things, to have any two planes intersect in a line. Try to find a Principle of Duality for the geometry you formulate. Explore some of the basic properties of this geometry. Is every plane in your geometry a projective plane in its own right?

2.3 AFFINE PLANES

In a projective plane any two lines contain a common point. In Euclidean plane geometry, it is not true that two lines need to contain a common point since it is quite possible for two lines to be parallel. In a projective plane we have no such things as parallel lines. If we wished to set up an axiom system for a linear space in which there was a parallel postulate such as the one included in Euclidean geometry, then we could use the following axioms.

Definition 5   *An **affine plane** is a linear space A having the properties:*

(6) *There are at least three noncollinear points of A.*

(7) *If a point P of A is not contained on a line L of A, then there is precisely one line L' which contains P and which does not share a common point with L.*

*If L and L' are lines of an affine plane which either are equal or have no common point, then we say that L is **parallel** to L'.*

Note that any line of an affine plane has been defined to be parallel to itself.

Example 4 Let A be the vertices of a square and let the lines of A be pairs of these vertices. Then A contains four points and six lines. It is easily verified that A is an affine plane. Note that unlike projective planes, affine planes can contain lines having two points. The example given here, however, is essentially the only example of an affine plane whose lines contain two points.

Example 5 Consider Example 1 again. Remove the points of any line L from π and label the resulting set $A(\pi)$. Then $A(\pi)$ contains four points. Let a line of $A(\pi)$ be a line of π less the point removed when L was taken away. Then $A(\pi)$ has six lines of two points each. It is easily verified that $A(\pi)$ forms an affine plane isomorphic to that proposed in Example 4.

In Example 5 we found an affine plane related to a projective plane. We now show that related to every projective plane, there is an affine plane.

Proposition 11 Let π be a projective plane and $\bar{L}$ be any line of π. We form a new structure $A(\pi, \bar{L})$ from π as follows: Let $A(\pi, \bar{L}) = \pi - \bar{L}$. The lines of $A(\pi, \bar{L})$ are the nonempty subsets of $A(\pi, \bar{L})$ of the form $A(\pi, \bar{L}) \cap L$, where L is a line in π. Then $A(\pi, \bar{L})$ with the linear structure described is an affine plane.

Proof We must verify that $A(\pi, \bar{L})$ is a linear space with the additional properties stated in Definition 5. We now give each of the properties that $A(\pi, \bar{L})$ must satisfy in order to be an affine space and supply a proof for each property.

i) Two distinct points of $A(\pi, \bar{L})$ are contained in a unique line: Let P and Q be distinct points of $A(\pi, \bar{L})$; then P and Q are distinct points of π. Therefore P and Q are contained in a unique line PQ of π. Therefore the unique line of $A(\pi, \bar{L})$ which contains P and Q is $A(\pi, \bar{L}) \cap PQ$.

ii) Any line of $A(\pi, \bar{L})$ contains at least two points: any line of π contains at least three points. In removing $\bar{L}$, we remove exactly one point from each line of π except $\bar{L}$. It follows, then, that each line of $A(\pi, \bar{L})$ contains one point less than each line of π; hence each line of $A(\pi, \bar{L})$ contains at least two points.

iii) There are three noncollinear points. We leave the proof of this fact as an exercise.

iv) If L' is a line of $A(\pi, \bar{L})$ and P is a point of $A(\pi, \bar{L})$ which does not lie on L', then there is a unique line of $A(\pi, \bar{L})$ which contains P but which does not share a point in common with L': the line L' of $A(\pi, \bar{L})$ comes from a line L of S. In fact, $L' = L - (L \cap \bar{L})$, if $L \cap \bar{L}$ is a point Q of π. Consider the line PQ of π. Now $A(\pi, \bar{L}) \cap PQ = PQ - \{Q\}$ is a line of $A(\pi, \bar{L})$ which contains P. Since L and PQ meet only in Q, the corresponding lines $PQ - \{Q\}$ and L' of $A(\pi, \bar{L})$ do not share any points; hence they are parallel. We have therefore shown that there is some line of $A(\pi, \bar{L})$ which contains P and is parallel to L'. We leave it as an exercise to show that $PQ - \{Q\}$ is the only line containing P which is parallel to L'.

Corollary *If the projective plane π is of order n, then $A(\pi, \bar{L})$ contains n^2 points and each line of $A(\pi, \bar{L})$ contains n points; moreover, there are $n^2 + n = n(n + 1)$ lines.*

Proof π contains $n^2 + n + 1$ points and we are removing exactly $n + 1$ of these points (that is, we take away the number of points on $\bar{L}$) to form $A(\pi, \bar{L})$; therefore $A(\pi, \bar{L})$ contains n^2 points. During the course of proving (ii) we noted that each line of $A(\pi, \bar{L})$ contains one less point than the corresponding line of π; hence each line of $A(\pi, \bar{L})$ contains n points. There is one less line in $A(\pi, \bar{L})$ than in π; hence $A(\pi, \bar{L})$ contains $n^2 + n$ lines.

Definition 6 *The affine plane $A(\pi, \bar{L})$ derived from a projective plane by removing the line $\bar{L}$ is called the **affine plane associated with**, or **corresponding to**, π **with respect to $\bar{L}$**. We will continue to denote this affine plane by $A(\pi, \bar{L})$.*

Observe that $A(\pi, \bar{L})$ seems to depend not only on π, but also on $\bar{L}$. If we could be sure that given any two lines $\bar{L}$ and $\bar{L}'$ of π, then $A(\pi, \bar{L})$ and $A(\pi, \bar{L}')$ are isomorphic, then we would be justified in simply writing $A(\pi)$ and speaking of *the* affine plane associated with π. However, the facts at hand so far about projective planes do not justify this assertion.

Also note that the fact that an affine plane can contain a different number of points and lines shows that the Principle of Duality does not apply to affine planes.

We will soon show that not only is there at least one affine plane associated with each projective plane, but also there is a projective plane associated with each affine plane. Before proving this important result, however, we must prove some preliminary results.

Proposition 12 *Any two lines of an affine plane contain the same number of points.*

Proof Let L and L' be distinct lines of the affine plane A. Let P be any point contained in L but not in L'. Such a point P must exist. For if P did not exist, then L would be a subset of L' from which we would have two distinct lines containing some two points. Similarly, let P' be any point which is contained in L' but not in L. Set $f(P) = P'$. Let X be any point of L. Then there is a unique line L_X which contains X and is parallel to PP' (Fig. 2.7). Now L_X must contain some point Y of L'. For if L_X were parallel to L', then PP' and L' would be distinct lines which contain P' and

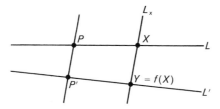

Fig. 2.7

are parallel to L_X, a contradiction of (7). Set $f(X) = Y$. It is left as an exercise to prove that the function f thus defined from L into L' is one-one and onto, and hence L and L' have the same number of points.

Proposition 13 *Let A be any affine plane in which each line contains exactly n points. Then there are precisely n^2 points, $n^2 + n$ lines, and given any line L, n lines of A which are parallel to L (Fig. 2.8).*

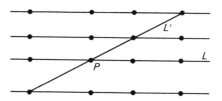

Fig. 2.8

Proof Suppose L is any line of A. Let L' be any line which shares a point P in common with L, but is different from L. (The existence of L follows from (6) in Definition 5 and is left to the reader to prove.) The lines L and L' each contain n points. For each of the n points of L' we have a line containing the point and parallel to L; moreover, all of these lines are distinct. Furthermore, if H is any line parallel to L, then H will contain some point of L'; if H did not, then L and L' would both contain P and be parallel to H, a contradiction of (7). Therefore there are n lines parallel to L (one for each of the n points of L').

Now every point of A is contained in one and only one of these lines. For if X is any point of A, there is a unique line parallel to L which contains X and the lines must intersect L'. From this it follows that each point of A is contained in one of the n lines which contain a point of L' and are parallel to L. Each of these lines contains n points, and no two of these lines contain a common point. We therefore conclude that there is a total of n^2 points in A.

We compute the number of lines in A as follows. Let L be any line and L' be any line parallel to, but not equal to, L. For each point X of L and each point Y of L', we have a distinct line XY (Fig. 2.9). There are n^2 such lines. The only lines left uncounted after all lines of the type XY have been

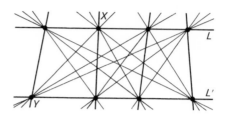

Fig. 2.9

formed are those parallel to L. There are, however, n such lines. Thus, there are $n^2 + n$ lines.

Definition 7 *The relationship "is parallel to" defines an equivalence relation of the set of lines of any affine plane* (Exercise 3). *If L is a line of an affine plane A, then the equivalence class of L* (*that is, the set of lines parallel to L*) *is said to be the **parallel class** of L. We denote the parallel class of L by* |L|.

Corollary *If A is an affine plane such that each line of A contains n points, then each parallel class contains n lines and there are n + 1 parallel classes.*

Proof That there are n lines in each parallel class follows at once from Definition 7 and Proposition 13. Since there are $n^2 + n$ lines and these are partitioned into parallel classes of n members each, there are

$$\frac{n^2 + n}{n} = n + 1$$

parallel classes.

The original inspiration for the projective plane was the phenomenon of perspective in our sense of vision. We note, for example, that parallel railroad tracks seem to meet at some great distance. If we had a large number of parallel lines side by side, they would all seem to meet, and meet in fact at the same point. The set of all "perspective points" at which the lines in parallel classes seem to meet can be thought of as a line itself; since this line appears to the eye to be at a great distance from us, we call it the *line at infinity*. Thus, any two parallel lines meet at a point on the line at infinity, and the line at infinity contains a point from each line.

The perceptive reader will remark that it appears to the eye that each line contains *two* points from the line at infinity since parallel lines appear to meet in both of two directions. We will, however, ignore the latter fact, even though this may seem inconsistent, and assume that there is but one point at which parallel lines meet on the line at infinity. The direction from which we view things does not matter. "Real" geometry without the lines at infinity gives the appearance of being affine (actually Euclidean), yet with the line at infinity we seem to obtain a projective geometry. Thus we seem to obtain a projective plane from an affine plane by adding another line in an appropriate way. We make these observations more precise in the following proposition.

Proposition 14 *Let A be any affine plane. Let $\bar{L}$ be any set having the same number of elements as there are parallel classes of A, and such that no element of $\bar{L}$ is a line or point of A.* (Thus there will be $n + 1$ elements of $\bar{L}$ if each line of A contains n points.) *Let $\pi(A) = A \cup \bar{L}$. To each parallel class of A we assign a unique element of $\bar{L}$ such that each element of $\bar{L}$ is assigned to one and only one parallel class. We denote the point assigned to $|L|$, the parallel class of L, by $P(|L|)$. The lines of $\pi(A)$ are defined as follows: Suppose P and Q are distinct points of $\pi(A)$. If P and Q are both contained in $\bar{L}$, set $PQ = \bar{L}$. Suppose P is contained in $\bar{L}$, but Q is in A. Then P has been assigned to a unique parallel class $|L|$. There is a unique line L' of $|L|$ which contains Q. Let $PQ = L' \cup \{P\}$. Suppose now that P and Q are both points of A. Then there is a unique line $(PQ)'$ of A which contains both P and Q. Set $PQ = (PQ)' \cup \{P(|(PQ)'|)\}$. Then $\pi(A)$ with the points and lines as described is a projective plane.*

Although the statement of Proposition 14 is lengthy, its proof is elementary and straightforward. We leave this proof as an exercise for the reader.

Definition 8 *The projective plane $\pi(A)$ of Proposition 14 is called the projective plane* **associated with,** *or* **corresponding to,** *A. $\bar{L}$ is called the* **line at infinity,** *while any point on $\bar{L}$ is said to be a* **point at infinity.**

The projective plane associated with an affine plane does not depend on the choice of the line at infinity (hence we are justified in talking about *the* projective plane associated with an affine plane), as we see from the following proposition.

Proposition 15 *If $\bar{L}'$ and $\bar{L}$ are any two sets which are used in accordance with Proposition 14 to form projective planes $\pi(A, \bar{L})$ and $\pi(A, \bar{L}')$ from an affine plane A, then $\pi(A, \bar{L})$ is isomorphic to $\pi(A, \bar{L}')$ by an isomorphism f for which $f(P) = P$ for each point P of A.*

Proof Set $f(P) = P$ for each point P of A. If Q is a point of $\bar{L}$, then $Q = P(|L|)$ for some parallel class $|L|$ of A. Let $P'(|L|)$ be the point of $\bar{L}'$ assigned to $|L|$ and set $f(P(|L|)) = P'(|L|)$. We let the reader verify that f is indeed an isomorphism from $\pi(A, \bar{L})$ onto $\pi(A, \bar{L}')$.

Propositions 11 and 14 essentially state that if we start with a projective plane, we can form an affine plane by omitting one line; and if we start with an affine plane, we can form a projective plane by adding a line. We can, therefore, draw conclusions about affine planes by studying projective planes, since any affine plane can be considered to be part of a unique projective plane. In truth, it is often much easier to study projective planes than affine planes because in projective planes we have the Principle of Duality, plus the simplifying condition that any two lines intersect. In an affine plane two lines might be either parallel or intersecting, and this leads to the necessity of considering many special cases.

EXERCISES

1. Prove Proposition 14.

2. Prove that the function f defined in the proof of Proposition 15 is an isomorphism.

3. Prove that "is parallel to" defines an equivalence relation of the set of lines of any affine plane.

4. The number of points in any line of an affine plane is called the *order* of the plane. Thus, an affine plane of order n gives rise to a projective plane of order n (Proposition 14), and a projective plane of order n gives an affine plane of order n (Proposition 11). Prove that if an affine plane has order at least 3, then it is generated by any three of its noncollinear points. Is this true of an affine plane of order 2?

5. Prove (iii) in the proof of Proposition 11. In (iv) of that same proof, show that $PQ - \{Q\}$ is in fact the only line which contains P and is parallel to L'.

6. Prove that the function f defined in the proof of Proposition 12 is one-one and onto as claimed.

7. Suppose that each line in an affine plane contains n points. How many lines contain any given point? Do this problem in two ways. First, compute this number directly from what we have proved about affine planes. Second, compute this number using what we know about projective planes. That is, use $\pi(A)$ to find the desired result about A.

8. Need there be an affine plane of order n for any integer $n \geq 2$? If the answer were affirmative, what could be said about the existence of projective planes of order n?

9. In the part of the proof of Proposition 13 that shows that A has $n^2 + n$ lines, prove that if L'' is any line not parallel to L, then L'' is not parallel to L'; hence $L'' = XY$ for some point X lying on L and some point Y lying on L'.

10. Suppose $\bar{L}$ and $\bar{L}'$ are two lines of a projective plane π. Try to find a necessary and sufficient condition for $A(\pi, \bar{L})$ to be isomorphic to $A(\pi, \bar{L}')$. Prove that $A(\pi(A), \bar{L})$ is isomorphic to A for any affine plane A, where $\bar{L}$ is the line at infinity. Is $\pi\big(A(\pi)\big)$ isomorphic to π?

11. Prove that given any point P of an affine plane A and L a line containing P, then there is a line L' distinct from L which also contains P.

2.4 PROJECTIVE SPACES

We conclude this chapter with a brief introduction to projective geometry of more than two dimensions. There are a number of reasonable ways to generalize the notion of a projective plane to higher-dimensional linear spaces. We might, for inspiration for a generalization, return to the phenomenon which originated the study of projective geometry, namely, perspective vision. Although we have restricted our attention previously to the projective plane, our field of vision is, of course, 3-dimensional. Two observations which we can make from our experience with vision are: first, any plane of vision has the properties of a projective plane; and second, any two planes appear to meet in a line. The first observation is one we might expect to hold in any projective space, regardless of its dimension; that is, it seems reasonable to expect that any 2-dimensional linear variety of a projective space should be a projective plane. The second observation is one we shall find characteristic of a projective space of three dimensions; for it can be shown that if we have a linear space S in which each 2-dimensional linear variety is a projective plane and in which any two distinct 2-dimensional linear varieties intersect in a line, then dim $S = 3$.

Definition 9 *A linear space having the property that any 2-dimensional linear variety is a projective plane is called a* **projective space**.

Proposition 16 *Any linear variety of a projective space is a projective space.*

Proof The proof is dependent on the lemma below. We leave the details of this proof, as well as the proof of the lemma, as an exercise for the reader.

Lemma *If S is any linear space and V is a linear variety in S, then V is a linear space in its own right with the geometry induced by S (cf. Example 5 of Chapter 1). Moreover, a subset W of V is a linear variety in V (that is, W is a linear variety relative to the geometry induced on V by S) if and only if W is a linear variety in S.*

Proposition 17 *If S is a projective space, then any two lines in S contain the same number of points.*

Proof Let L and L' be distinct lines of S. We distinguish two cases.

CASE 1 L and L' contain a common point P. Let Q be a point of L which is not in L', and Q' be a point of L' which is not in L. Then $V(\{P, Q, Q'\})$ is a 2-dimensional linear variety, and hence is a projective plane. Now L

and L' are both lines of $V(\{P, Q, Q'\})$. But any two lines of a projective plane contain the same number of points; hence L and L' contain the same number of points.

CASE 2 L and L' do not share a common point. Let P be any point of L and Q be any point of L'. Then by Case 1, L and L' each have the same number of points as PQ; hence L and L' have the same number of points.

The following definition helps in the statements and proofs of the next propositions.

Definition 10   *If A and B are two subsets of a linear space S, we define $\lambda(A, B) = \bigcup \{PQ \mid P$ is a point of A and Q is a point of $B\}$. If P is a point of S and A a subset of S, then $\lambda(P, A)$ is defined to be*

$$\bigcup \{PQ \mid Q \text{ is a point of } A\}.$$

Proposition 18   *Let S be a projective space, V a linear variety of S, and P a point of S which is not contained in V. Then*

$$V(V \cup \{P\}) = \lambda(P, V).$$

Proof   Certainly $\lambda(P, V)$ is a subset of $V(V \cup \{P\})$; it must be shown then that $V(V \cup \{P\})$ is a subset of $\lambda(P, V)$. This will be accomplished if we show that $\lambda(P, V)$ is a linear variety (and hence is in the family of linear varieties whose intersection is $V(V \cup \{P\})$). Let R and U be two distinct points of $\lambda(P, V)$; we will show that RU is contained in $\lambda(P, V)$. If R, U, and P are collinear, then RU is contained in $\lambda(P, V)$. For then R is contained in PQ for some point Q of V; hence

$$RU = PQ \subset \lambda(P, V).$$

Suppose now that R, U, and P are not collinear (Fig. 2.10). Then the lines PR and PU contain points R_1 and U_1 of V, respectively, and R_1 and U_1 are distinct. Consequently, P, R_1, and U_1 are noncollinear; hence $V(\{P, R_1, U_1\})$ is a projective plane. This projective plane, however, is easily seen to be

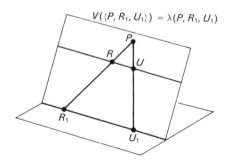

$$V(\{P, R_1, U_1\}) = \lambda(P, R_1, U_1)$$

Fig. 2.10

$\lambda(P, R_1 U_1)$. Since $R_1 U_1$ is contained in V,

$$V(\{P, R_1, U_1\}) = \lambda(P, R_1 U_1)$$

is contained in $\lambda(P, V)$. Since R and U are both points of $V(\{P, R_1, U_1\})$, RU is contained in $V(\{P, R_1, U_1\})$, and hence in $\lambda(P, V)$. Therefore $\lambda(P, V)$ is a linear variety and the proposition is proved.

Proposition 19 *If S is a projective space and L_1 and L_2 are two disjoint lines of S, then*

$$\lambda(L_1, L_2) = V(L_1 \cup L_2).$$

Proof Since $\lambda(L_1, L_2)$ is clearly a subset of $V(L_1 \cup L_2)$, it suffices to show that $\lambda(L_1, L_2)$ is itself a linear variety. Suppose that R and U are distinct points of $\lambda(L_1, L_2)$. By definition of $\lambda(L_1, L_2)$ there are points R_1 and U_1 on L_1 and R_2 and U_2 on L_2 such that R and U are contained in $R_1 R_2$ and $U_1 U_2$, respectively. If $R_1 = U_1$ (or if $R_2 = U_2$), then R and U are both points of $\lambda(R_1, L_2)$. We may then apply Proposition 18 to show that $\lambda(R_1, L_2)$ is a linear variety; hence we have

$$RU \subset \lambda(R_1, L_2) \subset \lambda(L_1, L_2).$$

We therefore assume that

$$R_1 \neq U_1 \qquad \text{and} \qquad R_2 \neq U_2.$$

Let X be any point on RU. Since R is a point of $R_1 R_2$ and U is a point of $\lambda(U_1, \lambda(R_1, L_2))$ as shown in Fig. 2.11, X is a point of $\lambda(U_1, \lambda(R_1, L_2))$, which is a linear variety by Proposition 18. Consequently, $U_1 X$ contains a point X' of $\lambda(R_1, L_2)$. Likewise $R_1 X'$ contains a point X'' of L_2; therefore the projective plane $V(\{R_1, U_1, X\})$ contains X' and X''. It follows, then, that XX'' and $L_1 = R_1 U_1$ share a common point Y. Therefore X is a point of $YX'' \subset \lambda(L_1, L_2)$. Consequently, RU is contained in $\lambda(L_1, L_2)$; hence $\lambda(L_1, L_2)$ is a linear variety.

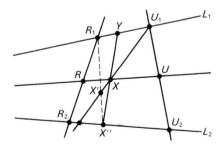

Fig. 2.11

Propositions 18 and 19 are special cases of the following proposition which, in turn, is proved by using Propositions 18 and 19. The actual proof is left as an exercise.

Proposition 20 *If V and V' are two linear varieties of a projective space S, then $V(V \cup V') = \lambda(V, V')$.*

We might suspect rightly that in a projective space of three dimensions there are no linear varieties of dimension greater than 2 which are properly contained in S. In such an instance, the 2-dimensional linear varieties of S would be *maximal* in the sense that no 2-dimensional linear variety is contained in a linear variety other than S of higher dimension. More generally, if a projective space has dimension n, then the maximal linear varieties of S will have dimension $n - 1$. This result may be entirely expected; yet it is not true of general linear spaces. In order to aid our discussion, we make the following definition.

Definition 11 *Let S be any linear space (not necessarily projective). A linear variety V of S, which is maximal in the sense that $V \neq S$, but if V' is any linear variety which contains V and $V' \neq S$, then $V = V'$, is called a* **hyperplane** *of S.*

Example 6 Each line in a projective plane π is a hyperplane. This follows at once from the fact that any three noncollinear points of π generate π. More generally, a line would be a hyperplane in any linear space generated by any three of its noncollinear points and having dimension at least two.

Example 7 Consider once more the linear space of Example 8 of Chapter 1. The space S itself has dimension 3 since A, B, and C generate it. We also note that $\{w, u, v\}$ is a linear variety in S which has dimension 3 but contains only three points. Even though dim $\{w, u, v\}$ = dim S, $\{w, u, v\}$ is a hyperplane since the only linear variety in S which contains it is all of S.

The next proposition gives a criterion for deciding if a linear variety of a projective space is in fact a hyperplane.

Proposition 21 *A linear variety V of a projective space S is a hyperplane if and only if each line of S contains at least one point of V.*

Proof Suppose first that every line contains at least one point of V. Then if P is some point not contained in V, we have $\lambda(P, V) = S$. Since any linear variety which contains V but is not equal to V must contain $\lambda(P, V)$ for some point P not in V, it follows that the only linear variety which properly contains V is S; hence V is a hyperplane.

Now suppose that V is a hyperplane. Let L be any line in S; we must show that L contains a point of V. If L did not contain a point of V, then

L would not be in $\lambda(P, V)$ for some point P outside of V (the verification of this statement is left to the reader). Therefore, $\lambda(P, V)$ would be a linear variety of S "between" V and S, contradicting the fact that V is a hyperplane.

Thus far what we have proved in this section is applicable to projective spaces of any dimension. We now look more closely at 3-dimensional projective spaces.

We saw earlier that any three points of a projective plane generate the plane. We have an analogous result for 3-dimensional projective space.

Proposition 22 *If S is a 3-dimensional projective space, then any four points of S which are not contained in some 2-dimensional linear variety of S generate S.*

Proof Since S is 3-dimensional, there are four points P, Q, R, and U which generate S. Suppose P', Q', R', and U' are four points of S which are not contained in some 2-dimensional linear variety of S. It must be shown that P, Q, R, and U are points of the linear variety generated by P', Q', R', and U'. This latter variety has the form $\lambda(P', \lambda(Q', R'U'))$. We leave the details of the proof to the reader.

The reader must beware of interjecting hypotheses into the proof which are not justified either directly by the definition of a projective space or by what we have already proved about projective planes and spaces. For example, Proposition 22 would follow quickly from the fact that 2-dimensional linear varieties are hyperplanes of S; yet this fact is a *corollary* of Proposition 22 and cannot be presumed in the proof (unless it is first proved independently of Proposition 22).

The following facts are almost immediate corollaries of Proposition 22; we leave their proofs to the reader.

Corollary 1 *The only linear varieties of a projective 3-dimensional space are the empty set, one point subsets, lines, and 2-dimensional linear varieties (projective planes), as well as S itself.*

Corollary 2 *If S is a 3-dimensional projective space, then any 2-dimensional linear variety is a hyperplane. Moreover, any 2-dimensional linear variety and any line share at least one point in common.*

Definition 12 *A 2-dimensional linear variety of a linear space is called a* **plane**.

Proposition 23 *Any two distinct planes of a projective 3-dimensional space S intersect in a line.*

Proof Suppose V and V' are distinct planes which do not intersect in a line. Since any line in V contains a point of V', it follows that $V \cap V'$

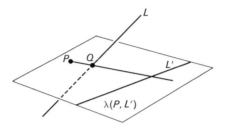

Fig. 2.12

consists of a single point P. Moreover, every line in V' must contain P (since every line of V' meets V and can do so only at P). Let L_1 and L_2 be distinct lines of V' and Q and Q' be points of L_1 and L_2, respectively, but different from P. Then QQ' is a line of V which does not contain P (the proof of this fact is left to the reader). This contradiction stemmed from the assumption that V and V' did not share a line in common; hence $V \cap V'$ is a line.

Proposition 24 *If S is a 3-dimensional projective space, and L and L' are lines of S which share no common point, and P is a point not contained in either L or L', then there is exactly one line in S which contains P and intersects L and L'.*

Proof The linear variety $\lambda(P, L')$ is a projective plane. Now L could not be a line of $\lambda(P, L')$ since then it would have to share a common point with L'. Consequently, L meets $\lambda(P, L')$ in some point Q (Fig. 2.12). Then the line PQ contains P and meets both L and L'. We leave it as an exercise to prove that PQ is the only line with this property.

This section has primarily been intended to give the reader some idea of projective geometry of higher dimensions. We will, however, use some of the results pertaining to 3-dimensional projective spaces in the next chapter.

EXERCISES

1. Prove Proposition 22.

2. Prove the corollaries to Proposition 22.

3. Complete the proof of Proposition 24.

4. Prove Proposition 20.

5. Let V be an n-dimensional vector space over a field F with $n \geq 3$. Let S be the set of 1-dimensional subspaces of V. If P and Q are points of S, let PQ consist of the points of S in the 2-dimensional subspaces generated by P and Q. (This is the same construction used in Example 2.) Prove that S, with the points and lines so defined, is a projective space. What is the dimension of S? Prove that the hyperplanes of S are sets of points of the $(n - 1)$-dimensional subspaces of S.

6. Formulate axioms for a 3-dimensional affine space. You should be able to prove that a 3-dimensional affine space can be obtained from a projective 3-dimensional space by removing a plane (analogous to the way that an affine plane was formed from a projective plane by removing a line). Can you get a projective 3-dimensional space from your affine space by "adding" a plane?

7. Suppose S is a projective space, H is a hyperplane of S and V is any linear variety of S which is not a subset of H. Prove that $H \cap V$ is a hyperplane of V with the geometry induced by S.

8. Prove that any two hyperplanes of a projective space are isomorphic. [*Hint:* Let P be a point of S not in H or H', the hyperplanes. For each X in each set, $f(X)$ is equal to the point of intersection of PX and H'.]

9. Prove that Proposition 22 is equivalent to: The hyperplanes of a 3-dimensional projective space are the planes of S.

10. Find the number of points in a 3-dimensional projective space S in which each line contains n points. Find the number of lines of S. Find the number of lines which contain each point. Can you state a Principle of Duality for projective 3-dimensional spaces?

11. In the proof of Proposition 21, prove that if L does not contain a point of V, then L would not be in $\lambda(P, V)$ for some point P outside of V.

12. Find the hyperplane of the linear space of Example 8 of Chapter 1.

13. Prove that if S is a projective space in which any two planes intersect in a line, then dim $S = 3$. This, together with Proposition 23, gives: A projective space S has dimension 3 if and only if any two planes of S intersect in a line.

3 Collineations of Affine and Projective Planes

3.1 COLLINEATIONS OF AFFINE PLANES

In Chapter 1 we saw that the set $G(S)$ of collineations of a linear space S, that is, the set of all one-one mappings of S onto itself which take lines into lines, is a group with composition of functions as the group operation. It is, of course, evident that the group structure of $G(S)$ will depend on the geometric structure of S. It is our purpose in this chapter to investigate certain of the properties of $G(S)$ for projective and affine planes. It would be impossible to give anything more than a glimpse of this important area of study in the space we can devote to it; nevertheless, we hope to raise some interesting questions, and give the reader a taste of the techniques and results one might expect to encounter in such a study.

We begin by looking at an affine plane A with collineation group $G(A)$, $\circ$. Our first proposition, however, applies to any linear space.

Definition 1 *A collineation f of a linear space S is said to fix a point x of S if $f(x) = x$. We say that f **fixes a subset** T of S if f fixes every point in T. We say that f **preserves** a subset T of S if $f(T) = T$ (even if f does not fix every point of T).*

We will denote the set of collineations of S which fix a subset T of S by $G(S: T)$.

Example 1 The reflection of the coordinate plane R^2 about the x-axis described by $(x, y) = (x, -y)$ fixes all points with y-coordinate 0, that is, this reflection fixes the x-axis. The function $f: R^2 \rightarrow R^2$ defined by $f(x, y) = (2x, 2y)$ preserves the x-axis, that is, takes any point on the line with equation $y = 0$ onto a point of this same line, but does not fix each point on the x-axis.

Proposition 1 $G(S: T)$ *is a subgroup of $G(S)$ for any linear space S and subset T of S.*

Proof Since the identity collineation fixes every point of S, it fixes every point of T; hence the identity collineation is in $G(S: T)$. Suppose f and g are members of $G(S: T)$. Then for each $t \in T$,

$$(f \circ g)(t) = f(g(t)) = f(t) = t;$$

hence $f \circ g$ is also in $G(S: T)$. Suppose $f \in G(S: T)$ and $t \in T$. Then if $f^{-1}(t) \neq t$, we have

$$f(f^{-1}(t)) = t \neq f(t),$$

which is not possible because f fixes T and $t \in T$. Therefore $f^{-1}(t) = t$ for each $t \in T$ and $f^{-1} \in G(S: T)$. Consequently, $G(S: T)$ is a group.

Proposition 2 *If f is a collineation of the affine plane A and if L and L' are parallel lines of A, then $f(L)$ and $f(L')$ are also parallel.*

Proof Since f is one-one and $L \cap L' = \varnothing$, we have $f(L) \cap f(L') = \varnothing$; hence $f(L)$ and $f(L')$ are parallel.

Corollary *If f is a collineation of the affine plane A, then $f(|L|) = |f(L)|$ for any line L of A. (By definition, $f(|L|)$ is*

$$\{f(L') \mid L' \text{ is parallel to } L\}.)$$

The corollary follows at once from Proposition 2.

In the next example, we consider collineations of the usual coordinate plane R^2 of analytic geometry. In a later section, we will generalize some of our considerations to "coordinate planes" over fields other than the real numbers.

Example 2 A collineation f of the coordinate plane R^2 with its usual geometry has the form

(1) $f(x, y) = (ax + by + c, a'x + b'y + c')$, a, b, c, a', b', c' real numbers, $ab' - a'b \neq 0$.

We make no attempt at the moment to justify this statement. All collineations, of course, take lines into lines. The collineation of (1) takes a line whose equation is $y = mx + q$ into a line whose equation can be readily determined in a straightforward but somewhat tedious fashion. We leave the finding of that equation to the reader. What we particularly want to discuss in this example is what sort of sets are fixed or preserved by various types of collineations of R^2.

i) There are collineations f of R^2 which have no fixed points whatsoever; that is, there is no point P of R^2 such that $f(P) = P$. Such collineations are called *translations*; translations have the general form

(2) $$(x, y) \rightarrow (x + c, y + c').$$

If we include the identity collineation in the set of translations, then the set of translations forms a subgroup of $G(R^2)$; moreover, given any points P and Q of R^2, there is a translation of R^2 which maps P onto Q. More formally, the group of translations of R^2 is transitive.

The translations have the property that they map any line L onto a line parallel to L; that is, $f(L)$ is parallel to L for any line L of R^2 and any translation f. (From an analytic geometric point of view, a translation maps a line of slope m onto a line of slope m. Although finding the equation of $f(L)$ in the case of a general collineation f is somewhat awkward, finding the slope of $f(L)$ for a translation f is a rather trivial matter. The reader should confirm that $f(L)$ has the same slope as L for any translation f and any line L.)

ii) It can be verified that any collineation having the form

(3) $$(x, y) \rightarrow (ax + c, ay + c')$$

also has the property of taking any line L into a line parallel to L. A collineation having this property is called a *dilatation*. Put in other words, a dilatation "fixes" parallel classes in the sense that for any dilatation f and any line L, $f(|L|) = |L|$ (cf. the corollary to Proposition 2).

iii) There are also collineations which have a fixed point, but no fixed parallel class. *Rotations* are examples of such collineations. For example, the collineation defined by

$$(x, y) \rightarrow \left(x/\sqrt{2} + y/\sqrt{2}, x/\sqrt{2} - y/\sqrt{2} \right)$$

is a rotation of $45°$ in a counterclockwise direction about the origin. the origin is the only fixed point of this collineation and there are no fixed parallel classes.

Although we could continue with our analysis of the various types of collineations of R^2, we assume that the reader now has some idea of what it is we are interested in examining, and we will, therefore, return to the study of general affine planes.

Definition 2 *Let A be any affine plane. A **dilatation** of A is a collineation of A which takes any line L onto a line parallel to L.*

*A **homothety with center C** is a dilatation which has C as a fixed point.*

*A **translation** is a dilatation which either is the identity collineation or has no fixed points at all.*

*A collineation f which fixes some line L (that is, $f(P) = P$ for all points P of L) is called an **affine perspectivity** with **axis** L.*

Example 3 A collineation of R^2 having the form

$$(4) \qquad\qquad (x, y) \rightarrow (ax, b'y)$$

is a homothety with center $(0, 0)$.

The collineation of R^2 defined by

$$(x, y) \rightarrow (-x, y)$$

is an affine perspectivity with the y-axis as its axis. (This would also be called a *reflection about the y-axis*.)

Proposition 3 *A homothety f with center C preserves any line which contains C.*

Proof Let L be any line which contains C. Then $f(L)$ is a line parallel to L and $f(L)$ contains $f(C) = C$. But the only line parallel to L which contains C is L itself; hence $f(L) = L$.

Proposition 4 *Let D(A) be the set of all dilatations of the affine plane A. Then D(A) is a normal subgroup of G(A).*

Proof We leave it as an exercise to show that $D(A)$ is in fact a subgroup of $G(A)$. In order to show that $D(A)$ is a normal subgroup, it suffices to show that if d is a dilatation and f is any collineation, then $f \circ d \circ f^{-1}$ is a dilatation. Let L be any line. Then $f^{-1}(L)$ and $d(f^{-1}(L)) = (d \circ f^{-1})(L)$ are parallel. Now any collineation takes parallel lines into parallel lines (Proposition 2); hence

$$f(f^{-1}(L)) = L \quad \text{is parallel to} \quad f((d \circ f^{-1})(L)) = (f \circ d \circ f^{-1})(L);$$

hence $f \circ d \circ f^{-1}$ takes L into a line parallel to L. Therefore $f \circ d \circ f^{-1}$ is a dilatation and the proposition is proved.

We know that a translation of the coordinate plane R^2, which carries the point P into the point $Q, P \neq Q$, preserves (but does not fix) the line PQ; indeed, such a translation preserves all lines parallel to PQ (Fig. 3.1). We now prove an analogous result for general affine planes.

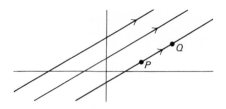

Fig. 3.1 A translation "moves" R^2 along the "direction" PQ, taking P onto Q.

Proposition 5 *Suppose f is a translation of the affine plane A, but f is not the identity collineation. Then for any point P of A, $f\big(Pf(P)\big) = Pf(P)$. Moreover, the set of all lines preserved by f forms a parallel class of A.*

Proof The line $Pf(P)$ is mapped by f onto the line $f(P)f^2(P)$. But $Pf(P)$ is parallel to $f(P)f^2(P)$ and both of these lines contain the point $f(P)$. Therefore $Pf(P) = f(P)f^2(P) = f\big(Pf(P)\big)$. Suppose now that L and L' are both lines such that $f(L) = L$ and $f(L') = L'$; we want to show that L is parallel to L'. If L and L' are not parallel, then they share a common point Q. Now $f(Q)$ is the point of intersection of $f(L) = L$ and $f(L') = L'$; hence it follows that $f(Q) = Q$. By assumption, however, f has no fixed points. Therefore L and L' must be parallel.

Suppose now that L' is parallel to L, and P is any point of L'. Then $f(P)$ is a point of $f(L')$. Since L' is parallel to L, $f(L')$ is also parallel to L. Therefore $Pf(P)$ and $f(L')$ are both lines which contain $f(P)$ and are parallel to L. It follows then that $Pf(P) = f(L')$. But since $Pf(P)$ and L' are both lines which contain P and are parallel to L, we have $L' = Pf(P) = f(L')$. Therefore f preserves L'.

We know that in the coordinate plane R^2, given points P and Q, there is always a translation which maps P onto Q; there is, in fact, only one such translation as we will soon see. We can also show that, given any line L of R^2, there is an affine perspectivity of R^2 with L as its axis. With regard to general affine planes, we want to answer the following type of question.

(5) Given two subsets T and T' of an affine plane A such that T and T' have the same number of elements, does there exist a certain type of collineation of A which maps T onto T'?

Specific examples of Question (5) would be: Given two lines L and L' of A, is there a dilatation (or translation, or affine perspectivity) of A which maps L onto L'? Given two points P and Q of A, is there a translation f of A with $f(P) = Q$? Given noncollinear points P, Q, and R, and noncollinear points P', Q', and R', is there a collineation f with $f(P) = P'$, $f(Q) = Q'$, and $f(R) = R'$?

The answer to the above questions is sometimes yes and sometimes no. It should be evident that the more collineations an affine plane has, the more likely the answer to any of these questions will be affirmative. For example, if the only collineation of A is the identity collineation, then the answer to the first question in the above paragraph is yes, if and only if $L = L'$. On the other hand, if every permutation of A is a collineation, then the answer will be yes with much less stringent restrictions.

Definition 3 *An affine plane A is said to have a **complete set of dilatations** if, given any distinct collinear points C, P, and Q, there is a homothety f of A with center C such that $f(P) = Q$, and given any points P and Q of A, there is a translation of A which maps P onto Q.*

We put the condition that C, P, and Q are collinear in Definition 3 because from Proposition 3 we have that any homothety with center C preserves the line CP for any point $P \neq C$; hence $f(P) = Q$ must lie on CP.

In the next section, we will investigate at some length a special class of affine planes which have a complete set of dilatations; we will also produce, later on, an affine plane whose set of dilatations is incomplete. Later it will be shown that the existence of a complete set of dilatations is equivalent to an important geometric property known as Desargues' Theorem.

So far in this section we have concentrated our attention on dilatations. We close the section by investigating some of the properties of affine perspectivities.

Proposition 6 *If f is an affine perspectivity other than the identity collineation, then f has precisely one axis L and has no fixed points which are not contained in L.*

Proof Suppose P is a point (of A) which is not contained in the axis L of f, yet $f(P) = P$. Then if Q is any point of L, we have $f(PQ) = f(P)f(Q) = PQ$; hence f preserves any line which contains P and a point of L. The only line which contains P but does not intersect L is the line which contains P and is parallel to L. Since f preserves parallelism, f preserves every line through P. We now prove that this implies that f is the identity collineation.

If f is not the identity collineation, then $f(Q) = Q'$ with $Q \neq Q'$ for some point Q (Fig. 3.2). Since f preserves lines containing P, $f(PQ) = PQ' = PQ$; hence P, Q, and Q' are collinear. Let Y be some point of L which does not lie on PQ, and let L' be the unique line which contains P and is parallel to YQ. Now $f(L') = L'$, but since $Q \neq Q'$, $YQ \neq YQ' = f(YQ)$ (why?); hence YQ' is not parallel to $f(L')$ (since YQ is the unique line which contains Y and is parallel to L'), contradicting the fact that f takes parallel lines into parallel lines. Therefore f must be the identity collineation.

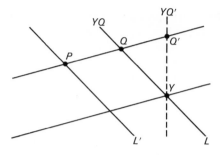

Fig. 3.2

Definition 4 *Suppose that f is an affine perspectivity of an affine plane A. Then for any point P of A for which f(P) ≠ P, the line Pf(P) is called a* **trace** *of f.*

Proposition 7 *Let f be an affine perspectivity other than the identity collineation, with axis L. Then for any point P, $f(Pf(P)) = Pf(P)$, provided $P ≠ f(P)$; that is, f preserves any of its traces. Moreover, the traces of f form a parallel class.*

Proof Suppose first that $Pf(P)$ contains a point Q of L. Then $Pf(P) = Qf(P)$. Hence $f(Pf(P)) = f(QP) = Qf(P) = Pf(P)$. On the other hand, if $Pf(P)$ shares no point in common with L, then it is parallel to L. But then $f(Pf(P)) = f(P)f^2(P)$ and $Pf(P)$ are both lines which are parallel to L and contain $f(P)$; hence they must be the same line. Thus, we again have $f(Pf(P)) = Pf(P)$.

It remains to be shown that all traces of f are parallel. Assume some trace L' meets L in some point R. Then all lines parallel to L' also meet L. For if L'' were parallel to L' and did not meet L, then L and L' would be two distinct lines which contain R and are parallel to L''. If L'' is a line parallel to L', then L'' contains a point U of L. Now $f(L'')$ is also a line parallel to $f(L') = L'$ which contains U; hence $f(L'') = L''$. Therefore L'' is a trace; hence all lines parallel to L' are traces. But each point P for which $P ≠ f(P)$ is a point of precisely one line parallel to L'; hence the only traces of f are the lines in $|L'|$.

If no trace intersects L, then the traces form the parallel class of L.

The proof of the following proposition is left as an exercise.

Proposition 8 *If L is a line, $|L'|$ a parallel class, and P and Q points of some line of $|L'|$ but not of L, then there is at most one affine perspectivity f with axis L, whose traces constitute $|L'|$, and $f(P) = Q$.*

Although Proposition 8 tells us that there is at most one affine perspectivity satisfying the special conditions stated in that proposition, it may be that there is no such affine perspectivity for certain L, $|L'|$, P, and Q. This leads to the following definition.

Definition 5 *An affine plane A is said to have a **complete set of affine perspectivities** if, given any L, $|L'|$, P, and Q as described in Proposition 8, there is an affine perspectivity f with axis L, $|L'|$ as its set of traces, and with $f(P) = Q$.*

The property of having a complete set of affine perspectivities, like the property of having a complete set of dilatations, is related to the Theorem of Desargues.

EXERCISES

1. Prove that any collineation of R^2 having the form (3) is a dilatation. Do rotations of R^2 fit into any of the categories of collineations of affine planes defined in Definition 2? Does "rotation" seem to be a meaningful concept for general affine planes? Explain. If you feel that the term is meaningful, try to give a definition of a *rotation* of an affine plane and explore some of the basic properties of rotations.

2. Prove that the set of all dilatations of an affine plane A is a subgroup of $G(A)$ (and thus complete the proof of Proposition 4).

3. Prove that the set of translations of an affine plane is a normal subgroup of the group of collineations $G(A)$. This proof may be broken down into the following steps.

   a) Prove that if f is a translation, then f^{-1} is also a translation.

   b) Prove that the composition of two translations is again a translation.

   c) Denote the set of translations of A by $T(A)$. Then (a), (b), and what is known about the identity collineation and composition of functions imply that $T(A)$ is a subgroup of $G(A)$. Suppose now that f is a translation and g is any collineation. Prove that $g \circ f \circ g^{-1}$ is a translation. You must show that $g \circ f \circ g^{-1}$ either has no fixed points or is the identity, and that $g \circ f \circ g^{-1}$ is a dilatation (this has already been done). We then have that $T(A)$ is a normal subgroup of $G(A)$.

   Prove that $g \circ f \circ g^{-1}$ preserves the parallel class $|g(P)g(f(P))|$.

4. Prove or disprove: The set of all homotheties with center C of an affine plane A is a normal subgroup of $G(A)$.

5. Prove that if P and Q are distinct points of an affine plane A, then there is at most one translation of A which maps P onto Q. (There may be no such translation.)

6. Answer the question "why?" which was posed in the last part of the proof of Proposition 6.

7. Does the set of affine perspectivities of an affine plane A form a subgroup of $G(A)$? Prove that all affine perspectivities with the same axis form a subgroup of $G(A)$. Is this a normal subgroup of $G(A)$?

8. Prove that a collineation of an affine plane A which preserves any two lines containing some point C is a homothety with center C. Prove or disprove: A collineation f of A, which has a fixed point C, is a homothety with center C.

9. Which of the various subgroups encountered thus far of the full group of collineations of an affine plane are commutative?

3.2 THE AFFINE PLANE OVER A FIELD

Thus far our consideration of affine and projective planes has been of a fairly abstract and general nature. We have, however, shown the reader how to construct a projective plane given any vector space of dimension 3 over any field whatsoever. This construction furnishes an endless number of projective planes for the reader to play with. Since an affine plane can be constructed from a projective plane, we have, at the same time, given the reader infinitely many examples of affine planes. Nevertheless, the projective planes obtained from vector spaces are rather abstract affairs, and the affine planes obtained from such projective spaces are even harder to visualize. We therefore pause a bit before continuing our abstract study and consider a very special and highly important class of affine planes. These affine planes are fairly easy to construct, and some are not hard to visualize; hence they may help the reader to digest some of the propositions already proved about affine planes. They will also give the reader some affine planes to work with as a basis for making new conjectures about general affine planes. Later in this section, we will relate the affine planes over fields with the projective planes defined, using vector spaces.

Definition 6 Let F be a commutative field*. Let

$$A(F) = F^2 = \{(x, y) \mid x \text{ and } y \text{ are elements of } F\}.$$

*Then $A(F)$ has the algebraic structure of a 2-dimensional vector space over F. We define a line in $A(F)$ to be any subset of $A(F)$ of the form $V + P = \{X + P \mid X \in V\}$, where P is any point of $A(F)$ and V is a 1-dimensional subspace of $A(F)$; that is, the lines of $A(F)$ are the cosets of 1-dimensional subspaces of $A(F)$. In Proposition 9, we will prove that $A(F)$ with lines so defined is an affine plane. We call this affine plane the **affine plane over the field F** and denote it by $A(F)$.*

---

* We actually do not need commutativity for most of what we are going to do, but we restrict ourselves to the commutative case for the sake of simplicity.

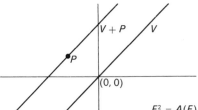

$F^2 = A(F)$ **Fig. 3.3**

Proposition 9 *With the structure described above, $A(F)$ is an affine plane. Moreover, the lines of $A(F)$ can be characterized as sets of points (x, y), satisfying equations of the form $y = mx + q$, or $x = q$, where m and q are elements of F.*

(The reader will observe that the affine plane $A(F)$ is nothing but a generalization of the usual analytic geometry on R^2; or, looking at it another way, the usual analytic geometry on R^2 is simply $A(R)$, a special case of what we have defined in Definition 6.)

Proof Let $P = (a, b)$ and $P' = (a', b')$ be distinct points of $A(F)$. Consider the line

$$PP' = \{t(a - a', b - b') \mid t \in F\} + (a', b').$$

PP' is the line since $\{t(a - a', b - b') \mid t \text{ in } F\}$ is a 1-dimensional subspace of $A(F)$ generated by $(a - a', b - b')$. Setting $t = 0$ and $t = 1$, we see that PP' contains P and P'. We leave it as an exercise to prove that PP' is the only line in $A(F)$ which contains both P and P'.

Since F contains at least two elements, each 1-dimensional subspace of $A(F)$ contains at least two elements; hence each line contains at least two points.

The points $(0, 0)$, $(1, 0)$, and $(0, 1)$ are easily shown to be noncollinear, the details being left as an easy exercise. Therefore there are three non-collinear points.

Suppose now that $L = V + (a, b)$ is any line in $A(F)$ and (a', b') is any point of $A(F)$ not contained in L. We will now show that $L' = V + (a', b')$ is parallel to L. Now L and L' are both cosets of the subgroup V of F^2. Any two cosets of a subgroup either coincide or are disjoint. Since L and L' do not coincide (since (a', b') is a point of L', but not of L), they must be disjoint. Therefore L and L' are parallel. That L' is the only line of $A(F)$ which contains (a', b') and is parallel to L will follow from the last part of Proposition 9.

Suppose $L = V + (a, b)$ and (c, d) generates V with $c \neq 0$. Then

$$L = \{(x, y) \mid x = tc + a \text{ and } y = td + b \text{ for } t \text{ in } F\}.$$

We can solve the equation $x = tc + a$ for t and substitute for t in $y = td + b$, obtaining

(6) $$y = c^{-1}dx - dac^{-1} + b.$$

We can verify that (x, y) is a point of L if and only if (x, y) satisfies (6); hence L has an equation of the form $y = mx + q$. If V has a generator of the form $(0, d)$, then L is characterized by the equation $x = a$. Therefore, any line has an equation of the form $y = mx + q$, or $x = q$. We now show that any solution set to an equation of the form $y = mx + q$, or $x = q$, is a line.

Suppose that $L = \{(x, y) \mid y = mx + q\}$, where m and q are elements of F. Then

$$L = \{t(1, m) \mid t \in F\} + (0, q),$$

which is a line. If $L = \{(x, y) \mid x = q\}$, then

$$L = \{t(0, 1) \mid t \in F\} + (q, 0),$$

which is also a line.

Now the lines whose equations are

$$y = mx + q \quad \text{and} \quad y = m'x + q'$$

are parallel if and only if $m = m'$. The proof of this fact is left as an exercise. If, then, $y = mx + q$ is the equation of some line, then any line parallel to it has an equation of the form $y = mx + q'$. If two lines containing (a', b') are parallel to the line with equation $y = mx + q$, then we have

$$b' = ma' + q' \quad \text{and} \quad b' = ma' + q''$$

for some q' and q'' of F. It follows that $q' = q''$; hence these lines must have the same equations, and therefore are the same line. The case for lines with equations of the form $x = q$ is left to the reader. This completes the proof of Proposition 9.

Certain collineations of $A(F)$ which we now introduce have special interest.

Definition 7   An *affinity* or *affine transformation* of the plane $A(F)$ is a function of the form

(7) $$(x, y) \rightarrow (ax + by + c, a'x + b'y + c'),$$

where $a, b, c, a', b',$ and c' are elements of F, and $ab' - ba' \neq 0$.

Proposition 10   An affinity of $A(F)$ is a collineation of $A(F)$. Moreover, the set of affinities of $A(F)$ forms a subgroup of the group of all collineations of $A(F)$.

Proof This proposition may be proved directly from (7) and the definition of a line in $A(F)$. We may also note, however, that the function (7) is the composition of a nonsingular linear transformation of F^2 into itself (the linear transformation whose matrix relative to the standard basis of F^2 is

$$\begin{pmatrix} a & b \\ a' & b' \end{pmatrix},$$

with the translation which takes $(0, 0)$ to (c, c')). Since each of these functions is a collineation of $A(F)$, their composition is also a collineation. Consequently, any affinity is a collineation.

One can verify directly from (7) that the composition of affinities is an affinity, or one can note that each affinity "factors" in a unique way into a nonsingular linear transformation composed with a translation. The group of affinities is isomorphic to the direct sum of the group of nonsingular linear transformations of F^2 with the group of translations.

It is possible for an affine plane $A(F)$ to have collineations which are not affinities, as we see from the following example.

Example 4 Let C be the field of complex numbers with $A(C)$ the affine plane on C. The *complex conjugate* of a complex number $a + bi$ is defined to be $a - bi$. The function j which takes each complex number into its complex conjugate is an automorphism of the field of complex numbers. It is left to the reader with some familiarity with complex numbers and linear algebra to show that the function from $A(C)$ into $A(C)$ defined by

$$(z, z') \rightarrow (j(z), j(z'))$$

is a collineation of $A(C)$, but is not an affinity.

The collineation of $A(C)$ given in Example 4 is but a special instance involving the following proposition, the proof of which we omit.

Proposition 11 *Let F be any field. If g is any automorphism of F, then a function from $A(F)$ into $A(F)$ of the form*

$$(x, y) \rightarrow (ag(x) + bg(y) + c, a'g(x) + b'g(y) + c'),$$

where $a, b, c, a', b',$ and c' are elements of F and $ab' - a'b \neq 0$, is a collineation of $A(F)$; moreover, any collineation of $A(F)$ has this form.

Corollary *Since the identity function is the only automorphism of R, the field of real numbers, all collineations of the affine plane $A(R) = R^2$, have the form*

$$(x, y) \rightarrow (ax + by + c, a'x + b'y + c'),$$

where a, b, c, a', b', c' are real numbers and $ab' - a'b \neq 0$. In other words, all collineations of the usual coordinate plane of analytic geometry are affinities.

The following proposition shows that the group of affinities of an affine plane $A(F)$ is rather "full." In particular, this proposition can be used to show that the sets of affine perspectivities and dilatations of $A(F)$ are both complete.

Proposition 12 *The group of affinities of the affine plane $A(F)$ is transitive with respect to triples of noncollinear points. More particularly, if P, Q, and R are three noncollinear points of $A(F)$ and P', Q', and R' are also three noncollinear points, then there is an affinity f of $A(F)$ such that $f(P) = P'$, $f(Q) = Q'$, and $f(R) = R'$. Moreover, this affinity is unique.*

Before proving Proposition 12, we prove a lemma.

Lemma *Three points $P = (a_1, a_2)$, $Q = (b_1, b_2)$, and $R = (c_1, c_2)$ of $A(F)$ are noncollinear if and only if*

$$D \equiv (b_1 - a_1)(c_2 - a_2) - (b_2 - a_2)(c_1 - a_1) \neq 0.$$

Proof If P, Q, and R are all contained in the line with equation $x = k$, then $D = 0 = b_1 - a_1 = c_1 - a_1$. If P, Q, and R are all contained in the line with equation $y = mx + q$, then $c_2 - a_2 = m(c_1 - a_1)$, where m can be computed to be $(b_2 - a_2)/(b_1 - a_1)$; hence again $D = 0$. Therefore if $D \neq 0$, then P, Q, and R are noncollinear.

Suppose now that $D = 0$. Let $b_1 \neq a_1$; then P, Q, and R are contained in the line whose equation is

$$y = (b_2 - a_2)(b_1 - a_1)^{-1}(x - a_1) + a_2.$$

If $D = 0$ and $b_1 = a_1$, then either $b_2 = a_2$, or $c_1 = a_1$. In this case, we have either $P = Q$, or all of the points lie on the line with equation $x = a_1$. Consequently, if $D = 0$, P, Q, and R are collinear. This completes the proof of the lemma.

Proof of Proposition 12 We first show that there is exactly one affinity f' such that $f'(0, 0) = (a_1, a_2)$, $f'(1, 0) = (b_1, b_2)$, and $f'(0, 1) = (c_1, c_2)$. The affinity f' would have the form

$$f'(x, y) = (ux + vy + w, u'x + v'y + w'),$$

where u, v, w, u', v', and w' are unknowns to be computed subject to the additional condition $uv' - vu' \neq 0$. It is easily verified that the only solutions for u, v, etc., giving f' the required properties, are $u = b_1 - a_1$, $v = c_1 - a_1$, $w = a_1$, $u' = b_2 - a_2$, $v' = c_2 - a_2$, and $w' = a_2$. It follows from Lemma 1 that $uv' - vu' \neq 0$ since P, Q, and R are noncollinear. If f' is an affinity which takes $(0, 0)$ to P, $(1, 0)$ to Q, and $(0, 1)$ to R, and g' is an affinity which takes $(0, 0)$ to P', $(1, 0)$ to Q', and $(0, 1)$ to R', then setting $f = g' \circ f'^{-1}$, we obtain an affinity which takes P to P', Q to Q', and R to R'.

Preliminary to stating the next proposition, we prove another lemma.

Lemma *If f is an affinity of A(F) with fixed points P and Q, then f fixes the line PQ, that is, $f(X) = X$ for each point X of PQ.*

Proof A point of PQ has the form $t(Q - P) + P = tQ + (1 - t)P$, where t is an element of F. It is easily verified that $f(tQ + (1 - t)P) = tf(Q) + (1 - t)f(P)$, which, in turn, is equal to $tQ + (1 - t)P$, since P and Q are fixed by f. Therefore f fixes PQ.

Proposition 13 *The affine plane $A(F)$ has a complete set of dilatations as well as a complete set of affine perspectivities. The group of affinities of $A(F)$ is also transitive with respect to the following sets of objects: (a) one-point subsets, (b) pairs of distinct points, (c) triples of noncollinear points, (d) lines, (e) couples of distinct nonparallel lines, and (f) couples of parallel lines.*

We leave the proof of Proposition 13 to the reader. There are, in addition to those mentioned above, other sets of objects for which the group of affinities of $A(F)$ is transitive, for example, triples of lines meeting in distinct points. The reader might try to find as many such sets as he can.

We conclude this section by investigating the relationship between the affine plane $A(F)$ and the projective plane $\pi(F^3)$. We begin this discussion by introducing the important notion of homogeneous coordinates.

The points of the projective plane $\pi(F^3)$ are the 1-dimensional subspaces of F^3. If (a, b, c) is any nonzero vector of F^3, then (a, b, c) generates a 1-dimensional subspace of F^3 which consists of all vectors of the form (ta, tb, tc), where t is an element of F. We can consider the points of $\pi(F^3)$ to be equivalence classes of nonzero vectors of F^3, where two nonzero vectors (a, b, c) and (a', b', c') are considered to be equivalent, if one is a scalar multiple of the other. In other words, each nonzero vector is allowed to represent the 1-dimensional subspace of F^3 which it generates. In this way, we introduce a type of *coordinatization* into $\pi(F^3)$, since ordered (nonzero) triples of elements of F now indicate points of $\pi(F^3)$ (just as ordered pairs of elements of F indicate points of $A(F)$); yet different ordered triples may designate the same point of $\pi(F^3)$. For example, $(1, 1, 1)$ and $(5, 5, 5)$ both represent the same point of $\pi(R^3)$, since $(1, 1, 1) = (1/5)(5, 5, 5)$. (Uniqueness can be introduced, if desired, in several ways. For example, we could demand that the first nonzero coordinate of an ordered triple used to designate a point of $\pi(F^3)$ be 1. Uniqueness of coordinates, however, is usually not required.)

Definition 8 *The nonzero elements of F^3, subject to the equivalence relation described in the paragraph above, are said to form a system of **homogeneous coordinates** for $\pi(F^3)$. We will adopt a fairly standard convention of using* capital letters such as A, B, C, and X_1, X_2, and X_3 to indicate the use of homogeneous coordinates.

The next proposition characterizes lines of $\pi(F^3)$ in terms of homogeneous coordinates.

Proposition 14 *A subset of $\pi(F^3)$ is a line if and only if it is the solution set to an equation of the form*

$$(8) \qquad AX_1 + BX_2 + CX_3 = 0,$$

where A, B, and C are elements of F, not all 0.

Proof It is a standard result in linear algebra that one equation in 3 unknowns (with some coefficient nonzero), such as (8), has a 2-dimensional subspace of F^3 as its solution set; hence the solution set is a line of $\pi(F^3)$. It will now be shown that any line of $\pi(F^3)$ has an equation of the form (8).

Consider the lines of $\pi(F^3)$ whose equations are

$$X_1 = 0, \qquad X_2 = 0, \qquad \text{and} \qquad X_3 = 0.$$

These three lines do not intersect in a common point since they contain only $(0, 0, 0)$ in common, but $(0, 0, 0)$ is not a point of $\pi(F^3)$. If L is any line other than one of these three lines, then L meets two of these lines in distinct points, say $(A, B, 0)$ and $(A', 0, C')$. If either B or C' is 0, then L would have to have the equation $X_2 = 0$, or $X_3 = 0$, respectively. If B and C' are both nonzero, then the line whose equation is

$$X_1 - AB^{-1}X_2 - A'C'^{-1}X_3 = 0$$

is a line which contains both $(A, B, 0)$ and $(A', 0, C')$; hence it must be the line L.

Before proceeding to discuss the relationship between $A(F)$ and $\pi(F^3)$, we digress a moment to see the effect of the Principle of Duality in the now coordinatized $\pi(F^3)$. Any nonzero ordered triple (A, B, C) of elements of F determines a unique point of $\pi(F^3)$; any nonzero scalar multiple of (A, B, C) determines the same point. Dually, (A, B, C) also determines a line in $\pi(F^3)$, namely, the line with equation $AX_1 + BX_2 + CX_3 = 0$; furthermore, any nonzero scalar multiple of (A, B, C) determines the same line. For convenience, when we consider (A, B, C) as the coordinates of a line, we will write $[A, B, C]$. The set of all points with homogeneous coordinates (X_1, X_2, X_3), which satisfy $AX_1 + BX_2 + CX_3 = 0$, is the line $[A, B, C]$. Dually, the set of all lines with (homogeneous) coordinates $[X_1, X_2, X_3]$, that is, those lines which satisfy $X_1A + X_2B + X_3C = 0$, consists of all lines which contain the point (A, B, C).

We now prove the basic theorem relating $\pi(F^3)$ and $A(F)$.

Proposition 15 Let $\pi(A(F))$ be the projective plane obtained from the affine plane $A(F)$ in accordance with Proposition 14 of Chapter 2. Then $\pi(A(F))$ is isomorphic to $\pi(F^3)$.

Proof The points of $A(F)$ are elements of F^2, while homogeneous coordinates are used to represent the points of $\pi(F^3)$. For any point (x, y) of $A(F)$, set

$$f(x, y) = (x, y, 1).$$

Recall that $\pi(A(F))$ was formed from $A(F)$ by adding a point $P(|L|)$ to each parallel class $|L|$ of $A(F)$. If a line L of $A(F)$ has the equation $y = mx + q$, then a line L' of $A(F)$ is in $|L|$ if and only if L' has an equation of the form $y = mx + q'$; thus, the parallel classes of lines with equations not of the form $x = q$ can be identified with their slopes m, which are elements of F. If $P(|L|)$ is the point associated with the parallel class of lines with slope m, we set

$$f(P(|L|)) = (1, m, 0).$$

For the point P' associated with the parallel class of lines with equations $x = c$, we set

(9) $$f(P') = (0, 1, 0).$$

We have now obtained a well-defined function f from $\pi(A(F))$ into $\pi(F^3)$. We must now show that f is one-one, onto, and takes lines into lines.

The function f is one-one and onto: Suppose $P = (X_1, X_2, X_3)$ is any point of $\pi(F^3)$. If $X_3 \neq 0$, then $P = (X_1 X_3^{-1}, X_2 X_3^{-1}, 1)$ and P is the image of the point $Q = (X_1 X_3^{-1}, X_2 X_3^{-1})$ of $A(F)$. From the definition of f, Q is the only point of $\pi(A(F))$ which is mapped by f into P. If $X_3 = 0$ and $X_1 \neq 0$, then $P = (1, X_1^{-1} X_2, 0)$, which is the image of $P(|L|)$, where $|L|$ is the parallel class of lines with slope $X_1^{-1} X_2$; furthermore, P is the image of this point alone. If $X_1 = X_3 = 0$, then X_2 must not be 0 and $P = (0, 1, 0)$ is the image of P' as in (9). Therefore f is one-one and onto.

The function f is linear: f takes the line at infinity of $\pi(A(F))$ into the line with equation $X_3 = 0$. A line whose equation in $A(F)$ is of the form $x = q$, together with its point at infinity, is carried by f onto the line with equation $X_1 = qX_3$. A line whose equation in $A(F)$ has the form $y = mx + q$ together with its point at infinity is carried by f onto the line with equation $X_2 = mX_1 + qX_3$. The reader is left to verify the truth of the assertions made concerning the action of f on lines. Therefore f is linear; hence f is an isomorphism from $\pi(A(F))$ onto $\pi(F^3)$.

EXERCISES

1. Do those parts of the proof of Proposition 9 which were left as exercises.

2. Prove that the set of affinities of an affine plane $A(F)$ is a group.

3. Prove that the function defined in Example 4 is a collineation of $A(C)$, but is not an affinity.

4. Prove that the only automorphism of the field of real numbers is the identity automorphism, thus establishing the corollary to Proposition 11.

5. Verify in the proof of Proposition 12 that the values of u, v, w, u', v', and w' given are the unique solutions to the given system of equations.

6. Prove Proposition 13.

7. Confirm the truth of the assertions about the action of f on lines made in the proof of Proposition 15.

8. Let F be the field Z_3 of integers modulo 3.
   a) Find all points on the line with equation $y = 2x + 1$ in $A(F)$.
   b) Find the equations of all lines of $\pi(F^3)$ which contain the point $(1, 0, 0)$. Do the same for all lines which contain the point $(2, 0, 1)$.
   c) Find all points of $\pi(F^3)$ which are contained in the line $[0, 1, 2]$ and all those which are contained in the line $[1, 1, 1]$.
   d) Find an affinity of $A(F)$ which takes the points $(1, 2)$, $(0, 0)$, and $(1, 0)$ onto the points $(1, 1)$, $(2, 1)$, and $(0, 0)$, respectively.
   e) Construct a "picture" of $\pi(F^3)$. Find $A(F)$ in this picture.

9. The corollary to Proposition 11 gives us the form for any collineation of the coordinate plane R^2. Try to characterize each of the various kinds of collineations of affine planes we have studied thus far in terms of the coefficients a, b, c, a', b', and c'. Partial results (without proofs) are indicated in Example 2 of this chapter.

10. Let F be any field and set $S = F^3$. Let the lines of S be subsets of S of the form $V + (a, b, c)$, where V is a 1-dimensional subspace of V and (a, b, c) is any point of F^3; that is, the lines of S are the cosets of 1-dimensional subspaces of F^3. Prove each of the following:
    a) S, with the structure described, is a linear space.
    b) The planes of S are subsets of the form $W + (a, b, c)$, where W is a 2-dimensional subspace of F^3 and (a, b, c) is a point of S.
    c) Any plane of S is an affine plane.
    d) Any two planes of S either fail to intersect or intersect in a line.
    e) If A is any plane of S and P is a point of S not contained in A, then there is exactly one plane which contains P and fails to meet A (that is, is parallel to A).
    f) S is a 3-dimensional, and each plane of S is a hyperplane.

    Since S has the properties (a) through (f), we can feel justified in calling S an *affine space* of dimension 3. How might this affine space S be "completed" so as to give a 3-dimensional projective space?

11. Let F be any field. Form a 3-dimensional projective space from F^4 in such a way that the nonzero elements of F^4 can be used as homogeneous coordinates for the projective space; denote the projective space by $PS(F^4)$. Try to find

the equations of the lines and planes of $PS(F^4)$. Find an isomorphism between $PS(F^4)$ and the projective space obtained by completing the affine space of Exercise 10.

3.3 COLLINEATIONS OF PROJECTIVE PLANES

Since any affine plane can be completed by the addition of a line to give a projective plane, we might expect the collineations of an affine plane A to be related to collineations of the corresponding projective plane $\pi(A)$. The following proposition gives the relationship.

Proposition 16 Let A be an affine plane and f a collineation of A. Then f can be extended in a unique fashion to a collineation f' of $\pi(A)$. That is, there is a unique collineation f' of $\pi(A)$ such that $f'(X) = f(X)$ for each point X of A.

Proof By definition, $\pi(A) = A \cup \bar{L}$, where $\bar{L}$ is the line at infinity (cf. Proposition 14 of Chapter 2). For each point X of A, set $f'(X) = X$. If $|L|$ is a parallel class of A, then $f(|L|)$ is also a parallel class of A; specifically, $f(|L|) = |f(L)|$. Set $f'(P(|L|)) = P(f(|L|))$. Then f' is a function from all of $\pi(A)$ into $\pi(A)$. We leave it as an easy exercise to prove that f' has the desired properties.

Actually Proposition 16 is a special instance of the following proposition.

Proposition 17 Suppose π and π' are projective planes, and L and L' are lines of π and π', respectively. Then $\pi - L$ and $\pi' - L'$, with the geometric structure induced from π and π', respectively, are affine planes. (These are the affine planes $A(\pi, L)$ and $A(\pi', L')$, respectively; cf. Proposition 11 of Chapter 2.) Then if f is an isomorphism from $\pi - L$ onto $\pi' - L'$, there is exactly one isomorphism f' from π onto π' such that $f'(X) = f(X)$ for each point X of $\pi - L$.

The proof of Proposition 17 is rather lengthy. The reader who is interested in a proof might see Theorem 2.22 of John Blattner's, *Projective Plane Geometry**. The reader should, in any event, prove Proposition 16, assuming Proposition 17.

As with collineations of an affine plane, we can distinguish various types of collineations of projective planes. One important method of obtaining collineations of projective planes can be motivated by considering a problem in the area which led to many of the early results in projective geometry, namely, painting.

Consider, as we did much earlier, the bottom rim of a lampshade. Assume that this rim is a perfect circle. This circle lies in a particular plane π of

---

* John Blattner, *Projective Plane Geometry*. San Francisco: Holden-Day, 1968.

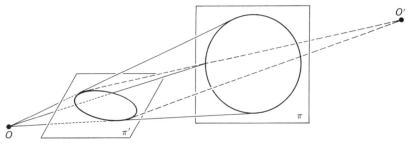

Fig. 3.4

real 3-dimensional projective space (considering the real space we live in as 3-dimensional and projective because that is the way we see things). The painter's canvas lies in another plane π' (Fig. 3.4). The painter wishes to place an image of the lampshade on his canvas, and to do so accurately so that his picture will be convincing. For the picture to be convincing, the scene on the canvas will have to appear just as the original scene appears to someone viewing it from a certain perspective. The proper way to draw the lampshade, if it is to appear on the plane π' as seen by someone looking at it from the point O, is to imagine the lampshade projected on the canvas in the manner suggested by Fig. 3.4. Formally, the projection of π on π', determined by the point O, gives an isomorphism of π onto π', with the image of the lampshade (with respect to this isomorphism) being the figure that the painter should draw. Note that the point O is fixed; it is from the perspective of O that the lampshade is viewed. If we now project π' on π, using the perspective of some point O', then the composition of this projection from O' with the original projection from O gives a collineation of π. By varying O and O', we can generate a large number of collineations of π. We will, in fact, use a method analogous to this in Section 3.4 to prove an important result.

The reader should note that projective collineations and isomorphisms need not preserve such properties as roundness; the image of the circle in Fig. 3.4 is not a circle, but an ellipse.

We now define an important type of collineation of a projective plane.

Definition 9 *Let π be a projective plane. A collineation f of π is said to be* **axial** *with* **axis** *L, if $f(X) = X$ for each point of the line L, that is, f fixes L.*

A collineation f of π is said to be **central** *with* **center** *C if f preserves all lines which contain C.* (Evidently, f will not fix all lines containing C unless f is the identity collineation.)

It appears at first that we have actually defined two types of collineations in Definition 9. Such is not really the case, as we see from the next proposition.

Proposition 18 *Any axial collineation is central, and any central collineation is axial.*

Proof Let f be a central collineation of a projective plane π. We must show that f fixes some line L of π. We leave it to the reader (see Exercise 3) to prove that any central collineation with two distinct centers is the identity collineation. If f is the identity collineation, then f is axial. Assume that f is not the identity and C is the unique center of f. Then C is a fixed point of f (why?). Suppose first that there is some line L such that $f(L) = L$, but C is not a point of L. Then f fixes L. For if Q is any point of L, then $f(CQ) = CQ$ and $f(L) = L$; hence Q, the intersection of L and CQ, is a fixed point of f. Therefore L serves as an axis for f and f is axial.

Suppose now that the only lines preserved by f are those which contain C. Let L' be any line which does not contain C; then $f(L') \neq L'$. Therefore, L' and $f(L')$ share a common point Q. Let $L = CQ$. Since L' and L are distinct lines, they share only the point Q in common. Now $f(Q)$ is a point of both $f(L) = L$ and $f(L')$; but Q is the only such point. Therefore Q is a fixed point of f. If X is any point of π other than Q for which $f(X) = X$, then $f(XQ) = XQ$; consequently, XQ must be a line which contains both C and Q; therefore, it is the line $CQ = L$. We have thus shown that every fixed point of f is contained in L. We now show that every point of L is fixed by f.

We already know that C is a fixed point of f. Suppose X is any point of L other than C. Let L'' be any line which contains X, but not C, and let Q' be the point common to L'' and $f(L'')$. Now $L'' \neq Q'C$ since L'' does not contain C; therefore, it follows that Q' is the point of intersection of L'' and $Q'C$. Consequently, $f(Q')$ is common to both $f(L'')$ and $f(Q'C) = Q'C$; hence $f(Q') = Q'$. Since all fixed points of f are contained in L, Q' is a point of L. Therefore Q' is a point of both L and L''; but this implies that Q' must be X. Therefore X is a fixed point of f; hence L is an axis for f and f is axial.

We now use the Principle of Duality to prove that any axial collineation is central. The statement we have already proved is: Any central collineation is axial. Restated, this says: Any collineation which preserves all of the lines which contain some point fixes all of the points of some line. The dual of this latter statement is: Any collineation which fixes all of the points of some line preserves all of the lines which contain some point. By the Principle of Duality we know that when a statement is true about a projective plane, its dual statement is also true. This completes the proof of Proposition 18.

Definition 10 *Let π be a projective plane. A central collineation (equivalently, an axial collineation) of π is said to be a **perspectivity**. (A perspectivity which is not the identity collineation has precisely one axis and one center.) If a* *perspectivity has its center on its axis, then we say that the perspectivity is a* **projective translation.** *If the center is not on the axis, we call the perspectivity a* **projective homothety.**

The following proposition relates the types of affine collineations we have studied, with their extensions to projective collineations in the completed affine plane.

Proposition 19 *Let A be an affine plane, $\pi(A)$ the associated projective plane, f a collineation of A, and f' the unique extension of f to $\pi(A)$ in accordance with Proposition 16. Then:*

a) *If f is a homothety, then f' is a projective homothety.*
b) *If f is a translation, then f' is a projective translation.*
c) *If f is an affine perspectivity whose axis is not parallel to a trace of f, then f' is a projective homothety.*
d) *If f is an affine perspectivity whose axis is parallel to a trace of f, then f is a projective translation.*

The proofs of these claims are straightforward and are left as exercises.

Proposition 20 *Let π be a projective plane, L a line of π, C a point, and P and Q points which are collinear with, but distinct from, C and are not contained in L. Then there is, at most, perspectivity f of π with axis L, center C, and $f(P) = Q$.*

Proof Suppose f and g are both perspectivities satisfying the given conditions. Then $f^{-1} \circ g$ is a central collineation with axis L, center C, and fixes P. If L' is any line which contains P, then L' contains two fixed points of $f^{-1} \circ g$, namely, P and the point of intersection of L and L'. Therefore $f^{-1} \circ g$ preserves L'. Consequently, P is a center for $f^{-1} \circ g$. But $f^{-1} \circ g$ is a perspectivity, and any perspectivity with two centers is the identity collineation. Therefore $f^{-1} \circ g$ is the identity; hence $g = (f^{-1})^{-1} = f$.

Proposition 20 tells us that there is, at most, one perspectivity satisfying the conditions set forth in its statement, but there may be no such perspectivity. As with affine planes, we are led to the following sort of definition.

Definition 11 *Let π be a projective plane. Then π is said to have a* **complete set of perspectivities** *if, given any line L and points C, P, and Q satisfying the conditions of Proposition 20, there is a perspectivity of π with axis L and center C, which maps P onto Q.*

As with affine planes, the existence of a complete set of perspectivities is related to the Theorem of Desargues. We will discuss this celebrated theorem and some of its consequences in the next section.

EXERCISES

1. Complete the proof of Proposition 16.

2. Assume Proposition 17 and prove Proposition 16 as a corollary.

3. Prove that any central collineation which has two distinct centers is the identity collineation. Prove or disprove: Any collineation of an *affine* plane which preserves the lines through two distinct points is the identity.

4. Any collineation of a projective plane which is the composition of perspectivities is called a *projectivity*. Prove or disprove each of the following assertions.
   a) The set of perspectivities of a projective plane is a group.
   b) The set of projectivities of any projective plane is a group.
   c) The set of perspectivities with center C is a group.
   d) The set of perspectivities with axis L is a group.
   e) The set of projective translations is a group.
   f) The set of projective homotheties is a group.

   Order these sets according to set inclusion; that is, which sets are subsets of which? Try to determine which of the groups are normal subgroups of the full group of collineations of the projective plane.

5. We have seen that any collineation of an affine plane A can be extended to a collineation of $\pi(A)$. Does every collineation of $\pi(A)$ correspond to a collineation of A? Might two distinct collineations of $\pi(A)$ correspond to the same collineation of A? Explain carefully your answers to the questions posed here. Does the set of collineations of $\pi(A)$ which correspond to collineations of A form a subgroup of $G(\pi(A))$? If so, is it a normal subgroup?

6. Let F be a (commutative) field. We have already considered an isomorphism between $\pi(F^3)$ and $\pi(A(F))$. Suppose f is an affinity of $A(F)$ with

   $$f(x, y) = (ax + by + c, a'x + b'y + c'), ab' - ba' \neq 0.$$

   Prove that the extension f' of f to $\pi(F^3)$, in accordance with Proposition 16, is defined by

   $$f'(X, Y, Z) = (aX + bY + cZ, a'X + b'Y + c'Z, Z).$$

7. Suppose f and g are central collineations of a projective plane π with centers C and C' and axes L and L', respectively. Prove that if C is a point of L' and C' is a point of L, then $f \circ g = g \circ f$.

8. Let $\pi(V)$ be the projective plane formed from the 3-dimensional vector space V over the field F. Prove that any nonsingular linear transformation of V onto V induces a collineation of $\pi(V)$. Find a necessary and sufficient condition for a collineation of $\pi(V)$ to be associated with a nonsingular linear transformation of V. Prove that $\pi(V)$ has a complete set of perspectivities.

3.4 COLLINEATIONS AND THE THEOREM OF DESARGUES

We stated earlier that there may or may not be some collineation of an affine or projective plane which satisfies certain given conditions. We are still faced with the question: Under what conditions do certain collineations exist? Clearly, for example, in order for a projective plane to have a complete set of perspectivities, it might have to have some special geometric property, but what property? It may even be (at least from what has been proved thus far in this text) that all projective planes have complete sets of perspectivities; we have not yet excluded this possibility (and will not do so until the next chapter). It may be that the only projective planes which have complete sets of perspectivities are those obtained from 3-dimensional vector spaces (cf. Exercise 8 of the preceding section). Such is, in fact, the case, although we will not give a complete proof of this in this section. We now introduce some notions which will help us to relate collineations of a plane (either affine or projective) and the geometric properties of the plane.

Since we have no way of designating line segments in a general affine or projective plane, we do not have triangles in the usual sense of Euclidean geometry. We can, however, define a generalized triangle.

*Definition 12 A **triangle** in a projective or affine plane is a set of three non-collinear points, together with the lines determined by these points. We will denote the triangle determined by the noncollinear points P, Q, and R by △PQR.*

Thus, if P, Q, and R are three noncollinear points of an affine or projective plane, then $\triangle PQR$ consists of P, Q, and R together with the lines PQ, QR, and PR (Fig. 3.5). Note that the dual statement of Definition 12 is: A "dual triangle" is a set of three lines which do not all contain the same point together with the points determined by these lines. In a projective plane, a dual triangle is nothing other than a triangle (is this true in an affine plane?). *Triangle*, then, is a *self-dual* notion since the dual of its definition is equivalent to the original definition. Another self-dual notion we have already encountered is *axial collineation*. The dual notion of *axial*

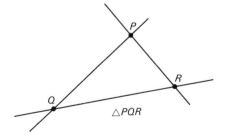

△PQR

Fig. 3.5

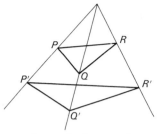

Perspective from a point   **Fig. 3.6**

collineation is *central collineation*, but a collineation of a projective plane is central if and only if it is axial.

Definition 13   *Two triangles are said to be **perspective from a point** if they can be labeled PQR and P'Q'R' in such a way that the lines PP', QQ', and RR' all meet in some point (Fig. 3.6)—or, in the case of an affine plane, they either meet in the same point, or are parallel.*

*Two triangles are said to be **perspective from a line** if they can be labeled ABC and A'B'C' in such a way that the points of intersection of corresponding sides, that is, of AB and A'B', BC and B'C', and AC and A'C' are collinear (Fig. 3.7)—or, in the case of an affine plane, the corresponding sides either intersect in collinear points, or are parallel.*

Observe that projective planes are easier to deal with, in regard to perspectivity from a point or a line, than is an affine plane, since in projective planes we do not have the possibility of nonintersecting lines. We are now ready to state Desargues' Theorem.

Proposition 21P (Desargues' Theorem for projective planes)   *Two triangles which are perspective from a point are perspective from a line.*

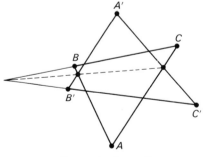

Perspective from a line   **Fig. 3.7**

Proposition 21A (Desargues' Theorem for affine planes)   *If two triangles PQR and P′Q′R′ are perspective from a point, and if PQ is parallel to P′Q′, and QR is parallel to Q′R′, then PR is parallel to P′R′.*

Definition 14   *A projective (affine) plane which satisfies Proposition 21P (Proposition 21A) is said to be **Desarguesian**.*

Looking just at the relative complexity of Propositions 21A and 21P, it appears the projective case will generally be easier to deal with than the affine case. The next proposition tells us that we can restrict our attention primarily to the simpler projective situation.

Proposition 22   *An affine plane A is Desarguesian if and only if π(A) is Desarguesian.*

Proof It is not particularly difficult to show that π(A) is Desarguesian, but we leave an indication of its proof until the next chapter.

Henceforth in this section, statements will be assumed to refer to a projective plane unless specifically stated otherwise. Before proving the basic result of this section, we prove a lemma.

Lemma   *Let f be a perspectivity with center C and axis L, and P be any point not on L and different from C. Then C, P, and f(P) are collinear. Moreover, if Q is any point of π, then f(Q) is uniquely determined once f(P) is known.*

Proof If f(P) is determined and is collinear with P and C, but different from C, then f is fully determined (Proposition 20). We will show explicitly, however, how f(Q) is determined, and, during this demonstration, it will be proved that C, P, and f(P) must be collinear. Assume then that f(P) is given; we may assume that Q is not P, C, or a point of L, since the action of f on all of these points is known. The line PQ meets L in some point U and f(U) = U (Fig. 3.8 or 3.9). Now f(PQ) = f(P)f(Q) contains U and f(P). We know that f preserves all lines which contain C. Since CP is such a line, f(P) must be a point of this line; hence P, C, and f(P) are collinear.

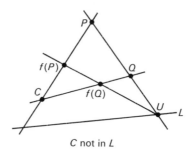

C not in L   **Fig. 3.8**

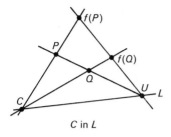

C in L

Fig. 3.9

Similarly, C, Q, and $f(Q)$ are collinear. Since $f(Q)$ is a point of both CQ and $Uf(P)$, it is the point of intersection of these two lines.

Corollary *If f is a perspectivity and L is a line which is not preserved by f, then the point of intersection of L and f(L) is a fixed point of f.*

Proof Let P be the point of intersection of L and $f(L)$. If $f(P) \neq P$, then the center C of f is collinear with P and $f(P)$. But then C, P, and $f(P)$ are all points of $f(L)$. This implies that L is a line which contains C; hence $f(L) = L$, contrary to hypothesis. Therefore P must be a fixed point of f.

Proposition 23 *A projective plane π is Desarguesian if and only if π has a complete set of perspectivities.*

Proof We first suppose that π has a complete set of perspectivities and $\triangle ABC$ and $\triangle A'B'C'$ are perspective from some point U. Let P, Q, and R be the points of intersection of CA and $C'A'$, BA and $B'A'$, and CB and $C'B'$, respectively (Fig. 3.10). We must show that P, Q, and R are collinear; that is, $\triangle ABC$ and $\triangle A'B'C'$ are also perspective from a line. Since π has

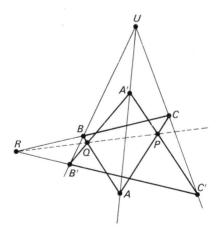

Fig. 3.10

a complete set of perspectivities, there is a perspectivity f with center U and axis PQ, such that $f(A) = A'$; moreover, this perspectivity is unique. Since P is a point of the axis of f, $f(P) = P$. Therefore $f(PA) = PA'$. Since U, C, and C' are collinear, and C is a point of PA while C' is on $Pf(A) = PA'$, it follows that $f(C) = C'$. Likewise, since Q is a fixed point of f, $f(QA) = QA$, from whence it follows that $f(B) = B'$. We therefore have $f(A) = A'$, $f(B) = B'$, and $f(C) = C'$. Consequently, $f(BC) = B'C'$; hence the point R is a fixed point of f by the corollary to the lemma. Therefore R is either the center U of f, or R is a point of PQ. If R is a point of PQ, then we have that triangles ABC and $A'B'C'$ are perspective from PQ. If $R = U$, we have that the line $UBB' =$ the line $RBB' =$ the line $RC'B' =$ the line $UB'C'$. Hence $UB = UC'$, which is impossible since $UA = UA'$, $UB = UB'$, and $UC = UC'$ are three distinct lines. Consequently, R must lie on PQ.

We now suppose that the projective plane π is Desarguesian and prove that π has a complete set of perspectivities. Let L be any line, U any point, and A and A' points which are collinear with U, but neither of which are contained in L. Our object is to find a perspectivity f with $f(A) = A'$ of which U is the center and L is the axis. Clearly we must set $f(X) = X$ for $X = U$ and for any point X of L. We now apply the construction given in the proof of the lemma preceding this proposition. Let B be any point of π which is not contained in L nor on $UA = AA'$; then AB meets L in some point Q and $A'Q$ meets UB in some point B'. We set $f(B) = B'$. This procedure defines $f(B)$ for any point B of π which is not contained in UA.

If C is some point not contained in L, with UC distinct from both UA and UB, then AC meets L in some point P, and PA' meets UC in some point C'. We set $f(C) = C'$. (We have simply carried out the procedure described above for finding $f(B)$, given B, to find $f(C)$, given C. Cf. Fig. 3.11.)

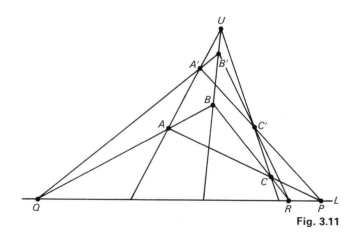

Fig. 3.11

We now define a function g in the same manner in which f was defined, except that we begin by setting $f(B) = B'$. Then g will be defined for all points of π, except those contained in UB. Given any point X of π, we can find either $f(X)$, $g(X)$, or both $f(X)$ and $g(X)$; hence if we show that f and g have the same definition for those points at which they both are defined, then we will have a function from all of π into itself. We apply Desargues' Theorem to the triangles ABC and $A'B'C'$ (Fig. 3.11). Since Q is a point of both AB and $A'B'$, and P is contained in both AC and $A'C'$, and both P and Q are points of L, we have R, the point of intersection of BC, and $B'C'$ is also a point of L. Therefore $B'R$ and UC have the point C' in common; it follows then that $g(C) = C'$, from whence we have $f(C) = g(C) = C'$. Therefore f and g agree wherever they are both defined. Consequently, we have a function from all of π into itself; we will denote this function by f. We leave it to the reader to show that f has the following properties: f fixes L, preserves all lines which contain U, and is one-one and onto. To complete the proof that f is a collineation with the desired properties, we must show that f is linear. The remainder of the proof is devoted to proving that f is linear.

Let L' be some line which does not contain U, and let A be some point not contained on L'; also let B and C be distinct points of L' different from the point P of intersection P of L and L' (Fig. 3.12). We define $A' = f(A)$, $B' = f(B)$, and $C' = f(C)$. Since P is a point of L, $f(P) = P$. It follows from the way that f was defined that Q and R, the points of intersection of AB and $A'B'$ and AC and $A'C'$, respectively, are both points of L (the reader should confirm this fact). Applying Desargues' Theorem to triangles $\triangle ABC$ and $\triangle A'B'C'$, we find that B', P, and C' are collinear. Therefore f maps the line $L' = PB$ onto the line PB'. There is one point at which the function g must be used, and that is for the point of intersection of PB and UA since f, as originally defined, is not defined for points of UA. It is easily verified, however, that the point of intersection of PB and UA is mapped by g onto the point of intersection of PB' and UA. This completes the proof of Proposition 23.

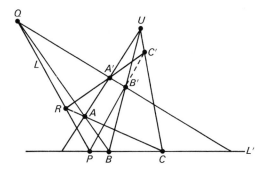

Fig. 3.12

We know that any plane of a projective space is a projective plane. We now prove that any plane of a projective 3-dimensional space must be Desarguesian.

Proposition 24 *If* π *is a hyperplane of a 3-dimensional projective space* S, *then* π *has a complete set of perspectivities.*

Proof Let L be a line, C a point, and P and Q points not contained in L, but collinear with C, all in π. Let R be some point of S which is not contained in π. Then R and L determine a plane π'. Let U be a point of RP different from both P and R. Then U is neither in π nor in π'. Now RP and Q determine a plane π'' and RQ is a line of this plane. Likewise, PQ is in π''; therefore C is in π''. Since U is a point of RP, U is in π''. Therefore CU is in π''. Consequently, CU and RQ contain a single point U' in common.

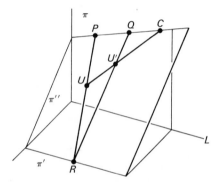

Fig. 3.13

Let h be the projection of π onto π' from the "perspective" of U (cf. the discussion following the proof of Proposition 17). That is, for each point X of π, UX intersects π' in a unique point; it is this point we call $h(X)$. Let h' be the projection of π' onto π from the perspective of U', and set $f = h' \circ h$. Then f is a collineation of π (since it is the composition of two isomorphisms). Moreover, f has been constructed so as to make it a perspectivity of π with center C, axis L, and with $f(P) = Q$.

Corollary 1 *If* π *is a projective plane which is isomorphic to a hyperplane* π' *of a projective 3-dimensional space* S, *then* π *has a complete set of perspectivities.*

Proof Let f be the isomorphism from π onto π', and L, C, P, and Q be a line and three points of π which might determine a perspectivity of π with axis L and center C. Let h be the perspectivity of π' with axis $f(L)$, center $f(C)$, and with $h(f(P)) = f(Q)$. Then $f^{-1} \circ h \circ f$ is the desired collineation of π.

Corollary 2 *If π is isomorphic to a hyperplane of a projective 3-dimensional space, then π is Desarguesian.*

Since not all projective planes are Desarguesian, there are projective planes which are not isomorphic to any plane of a projective 3-dimensional space.

Corollary 3 *Any projective plane of the form $\pi(F^3)$, where F is a field, is Desarguesian, and hence has a complete set of perspectivities.*

Proof We can form a projective 3-dimensional space of the form $PS(F^4)$ in a manner analogous to the way $\pi(F^3)$ was formed (cf. Exercise 11 of Section 3.2). We have homogeneous coordinates for $PS(F^4)$. It can be shown that the set

$$\{(X, Y, Z, 0) \mid X, Y, \text{ and } Z \text{ elements of } F \text{ which are not all } 0\}$$

is a hyperplane of $PS(F^4)$ (specifically, it is the hyperplane with equation $X_4 = 0$), and is isomorphic to $\pi(F^3)$ (by the isomorphism $(X, Y, Z) \leftrightarrow (X, Y, Z, 0)$; see Exercise 11). Consequently, $\pi(F^3)$ is isomorphic to a hyperplane of a projective 3-dimensional space and therefore is Desarguesian.

Corollary 4 *Any affine plane of the type $A(F)$, F is a field, is Desarguesian.*

Proof The plane $\pi\big(A(F)\big)$ is isomorphic to $\pi(F^3)$ which we have just seen is Desarguesian. By Proposition 22, then, $A(F)$ is Desarguesian.

Proposition 25 *Any affine plane A which is Desarguesian*
a) *has a complete set of dilatations, and*
b) *has a complete set of affine perspectivities.*

Proof We prove (a) and leave the proof of (b) as an exercise. Let A be any Desarguesian affine plane; then $\pi(A)$ is Desarguesian. Let $\overline{L}$ be the line at infinity of $\pi(A)$. Suppose C, P, and Q are distinct collinear points of A. Let f be the perspectivity of $\pi(A)$ which has $\overline{L}$ as its axis, C as its center, and $f(P) = Q$. Then f acts on A alone as a dilatation of A, with center C, and $f(P) = Q$.

Let g be the perspectivity of $\pi(A)$ with center $P(|PQ|)$, axis $\overline{L}$, and $f(P) = Q$. Then g acts on A as a translation, with $g(P) = Q$. Therefore, A has a complete set of dilatations.

Using Propositions 16, 19, and 22, we can also prove:

Proposition 26 *If an affine plane A has a complete set of dilatations, or a complete set of affine perspectivities, then A is Desarguesian.*

The details of this proof are left as an exercise.

Corollary *The following statements about an affine plane A are equivalent.*

a) *A is Desarguesian.*

b) *A has a complete set of dilatations.*

c) *A has a complete set of affine perspectivities.*

d) $\pi(A)$ *is Desarguesian.*

Observe that we have been able to obtain important results about affine planes by studying projective planes. The projective planes were more easily discussed because of the Principle of Duality and because there are no parallel lines in a projective plane (thus, we avoid annoying and time-consuming special cases).

EXERCISES

1. Prove that if A is an affine plane and $\pi(A)$ is Desarguesian, then so is A.

2. Prove each of the following in the second part of the proof of Proposition 23.
   a) Verify that f and g agree at each point, for they are both defined.
   b) Prove that f is one-one and onto.
   c) Prove that f preserves all lines which contain U.
   d) Prove that the points P, Q, and R in the proof that f is linear are all contained in L.
   e) Verify that g maps the point of intersection of PB and UA into the point of intersection of PB' and UA.

3. Prove that the projections defined in the proof of Proposition 24 are isomorphisms. Prove that the function f constructed in that proof is a perspectivity of π which has the properties claimed for it.

4. Supply the details for the proof of Corollary 3 of Proposition 24.

5. Prove (b) of Proposition 25.

6. Prove Desargues' Theorem for projective planes of the type $\pi(F^3)$ directly, using the analytic geometry of such planes.

7. Prove Proposition 26.

8. Suppose π is a Desarguesian projective plane and f and g are projective translations of π with a common axis L. Prove that $f \circ g = g \circ f$.

9. Prove that any 2-dimensional linear variety of any projective space of dimension at least 3 is Desarguesian.

10. A *projectivity* of a projective plane π is any composition of perspectivities. The following facts are of great importance, although limitations of time and space have prevented our discussing them in this text. The reader should try to prove them on his own. Failing this, for the proofs are rather difficult, the reader might look up proofs in another text.

a) If Desargues' Theorem holds in π, then any projectivity is the composition of, at most, two perspectivities.

b) Given noncollinear points P, Q, and R and noncollinear points P', Q', and R' of π, then there is exactly one projectivity f of π, such that $f(P) = P'$, $f(Q) = Q'$, and $f(R) = R'$.

11. Prove that the function defined in the proof of Corollary 3 to Proposition 24 is an isomorphism.

4 Some Algebraic Implications of Geometry

We have already seen that for any field F, $A(F)$, the affine plane on F, is very similar to the coordinate plane of ordinary real analytic geometry in that each point of $A(F)$ is uniquely determined by a point (x, y) of F^2, and each line of $A(F)$ has an equation either of the form $y = mx + q$, or $x = q$, where m and q are elements of F. In the case of standard analytic geometry, the field in question is the real number field. In sum, then, $A(F)$ is a "coordinate plane," and its linear structure can be specified analytically. We may, however, also ask: If an affine plane A has particularly nice geometric properties, must it, in fact, be isomorphic to an affine plane on some field? That is, given an affine plane A which has a suitable geometry, can we find a field F so that we can coordinatize A in such a way that from a set-theoretic and analytic-geometric point of view, A looks exactly like $A(F)$? In seeking to answer this question, we are clearly trying to relate geometric properties with algebraic properties; that is, we are trying to draw out some of the algebraic implications of the geometric structure.

If we merely view coordinates as labels to distinguish one point of a plane from another, then any plane, either projective or affine—indeed, any linear space—can be coordinatized. One need merely find a suitable set of labels which has the same number of elements as the plane and give each point of the plane a distinct label. Although we might, in such a fashion, technically be coordinatizing the plane—in the same way that putting coordinates on a map helps us to find a particular location more easily—we would not be relating the coordinate structure (that is, the labeling) with the linear structure of the plane. We want, rather, to assign coordinates so that lines can be given equations and so that analytic geometry is possible. The general method we will use to assign coordinates to the affine plane is due primarily to the eminent American mathematician Marshall Hall, Jr. This method was first published in a paper in the prestigious *Transactions of the American Mathematical Society* in 1943. Although that paper dealt with coordinatizing projective planes, the method is easily adapted for use with affine planes.

Let A be an affine plane. We select a point O from A (O will serve as the *origin*) and three distinct lines which contain O. (In any affine plane of order n, a point is contained in $n + 1$ lines; see Exercise 7, Section 2.3.) We call these lines the *x-axis*, the *y-axis*, and the *unit line* (Fig. 4.1). We have a great deal of latitude in our choices. Which point will be O, which three lines will contain O, and which of the lines selected will be the x-axis, y-axis, and unit line, are arbitrary choices; but once these choices are made, they must remain fixed throughout the process of coordinatization. (For example, we cannot decide on a new x-axis half-way through the coordinatization.)

We now need a set of elements with which to form coordinates. To this end we select a set K having the following two properties:

1. K contains two elements denoted by 0 and 1.
2. There are precisely as many elements of K as there are points on some line of A. (Any line will do since all lines contain the same number of points.)

To each point of the unit line we assign a unique ordered pair of the form (x, x), where x is an element of K. This assignment is carried out in such a way that each element of $K \times K$ of the form (x, x) determines a unique point of the unit line, and such that O is given the coordinates $(0, 0)$. Such an assignment is possible since $\{(x, x) \mid x \in K\} \subset K \times K$ has the same number of elements as K, which, in turn, has the same number of elements as the unit line.

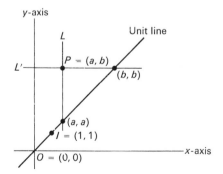

Fig. 4.1

Thus, O has the coordinates we would expect it to have (we have assigned the coordinates to make this so), and each point of the unit line has coordinates (a, a) for some element a of K. We call the point $(1, 1)$ of the unit line the *unit point* and denote it by I. Thus, the unit line is OI; the unit line is also the line whose equation is $y = x$, another expected result. We now have coordinates assigned to each point of the unit line; we shall proceed to assign coordinates to each point of A.

Let P be any point of A. There is a unique line L which contains P and is parallel to the y-axis; moreover, L intersects the unit line in some point (a, a). For if it did not, then the unit line and the y-axis would be distinct lines which contain O and are parallel to L. Similarly, there is a unique line L' which contains P and is parallel to the x-axis; L' intersects the unit line in some point (b, b). We let the *x-coordinate* of P be a, and the *y-coordinate* of P be b, giving P the *coordinates* (a, b) (Fig. 4.1). (If the reader carries out this procedure in R^2, with the usual x-axis, y-axis, and the line with equation $y = x$ as the unit line, he will obtain the usual coordinates. Our figure is drawn in such a way as to imply that the process looks much the same in any affine plane as it does in R^2. But the reader should keep in mind that not only is it difficult to obtain a graphical representation of most affine planes, but also in general affine planes we have no such things as angles, or perpendicularity, or many of the other niceties that Fig. 4.1 implies.)

The following proposition catalogs some of the obvious properties of the coordinates thus assigned.

Proposition 1

a) *Each point of the x-axis has coordinates of the form $(a, 0)$, and each point of the y-axis has coordinates of the form $(0, b)$, where a and b are elements of K.*

b) *Any line parallel to the y-axis has an equation of the form $x = a$, and any line parallel to the x-axis has an equation of the form $y = b$, where a and b are elements of K.*

c) *Each point of A is assigned precisely one element of $K \times K$, and each element of $K \times K$ determines a unique point of K.*

d) *The unit line has the equation $y = x$.*

So far we have equations for only a limited number of lines, even though the equations we do have are the expected ones. The ideal would be to have all other lines have some equation of the form $y = mx + q$, for some elements m and q of K. But such an equation implies the existence of algebraic operations (addition and multiplication) in K. However, K is just a set for which no algebraic structure has been assumed. If any algebraic operations are to be found, they must be found from A, not from K.

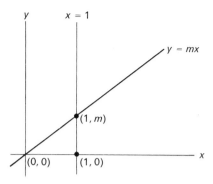

Fig. 4.2

In the ordinary coordinate plane R^2, if we take some line which contains the origin but is not the y-axis, then this line will intersect the line with equation $x = 1$ in some point $(1, m)$ (Fig. 4.2). We readily see that the given line then has equation $y = mx$; hence m is the slope of the given line, as well as the slope of any line parallel to it. We apply an analogous device now in A to assign slopes to lines.

Suppose L is any line in A which is not parallel to the y-axis. Let L' be the line which contains $(0, 0)$ and is parallel to L. Then L' intersects the line with equation $x = 1$ (this latter line is sometimes called the *slope line*) in some point $(1, m)$. We call m the *slope* of the line L.

The following proposition summarizes some of the obvious properties of the slope. The slope of lines parallel to the y-axis is left undefined.

Proposition 2

a) *If L and L' have the same slope, then they are parallel; and if two lines are parallel and the slope of one is defined, then the slopes of both lines are defined and are equal.*

b) *Lines which are parallel to the x-axis have slope 0.*

The reader may now be tempted to suppose that a line with slope m and y-intercept $(0, q)$ will have the equation $y = mx + q$. But, looking again, he will see that once more we are stymied by the fact that no algebraic operations have as yet been introduced. All we have done is introduce symbols and define things in such a way as to ensure that the expected will occur. We now introduce some algebra, not in the form of *binary operations*, that is, operations involving two elements, but in the form of a *ternary operation*.

If H is a set with a binary operation $\#$, this means that given two elements h and h' of H, $h \# h'$ is a uniquely determined element of H. (If $\#$ is a *commutative* operation, then $h \# h'$ will be the same as $h' \# h$, but not all operations are commutative. For example, composition of functions is not usually commutative.) Thus, a binary operation # on H is actually a function from $H \times H$ into H; that is, # assigns a unique element of H to each ordered pair of elements of H; instead of using function notation for #, that is, writing #(h, h'), we usually write h # h' to denote the image of (h, h') under #. A ternary operation F on H is simply a function from $H \times H \times H$, the set of ordered triples of H, into H. The example below illustrates a ternary operation.

Example 1 Let R be the field of real numbers with its ordinary addition and multiplication. Define

$$F(x, m, q) = mx + q,$$

for each ordered triple (x, m, q) of real numbers. Then F is a ternary operation on R. Moreover, each line of the coordinate plane R^2 which is not parallel to the y-axis has an equation of the form

$$y = F(x, m, q)$$

for some fixed real numbers m and q. We will soon define a ternary operation related to an affine plane which will make a similar statement true there.

Also note that addition and multiplication of real numbers can be defined in terms of the ternary operation F. Specifically, for any real numbers a and b,

$$a + b = F(a, 1, b)$$

and

$$ab = F(a, b, 0).$$

Once we have a ternary operation for an affine plane, we will be able to define binary operations corresponding to addition and multiplication. We now proceed to define a ternary operation F associated with the affine plane A we have been considering.

Given an ordered triple of elements of K, we want to specify $F(a, m, b)$. Consider the line L of slope m which passes through the point $(0, b)$ (Fig. 4.3).

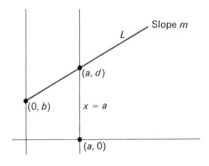

Fig. 4.3

Since L has a slope, L is not parallel to the y-axis; hence the line with equation $x = a$ must intersect L in a point whose coordinates are (a, d) for some element d of K. We set

(1) $$F(a, m, b) = d.$$

The following proposition summarizes some of the more obvious properties of the ternary operation F.

Proposition 3 *Any line with slope m and y-intercept $(0, q)$ has the equation*

$$y = F(x, m, q).$$

Corollary $x = F(x, 1, 0)$ *for each element x of K.*

Proof The unit line has equation $y = x$, but by Proposition 3 it also has the equation

$$y = F(x, 1, 0).$$

It follows then that

$$x = F(x, 1, 0).$$

Note that if we did define $ab = F(a, b, 0)$, then we would have the expected result $a \cdot 1 = F(a, 1, 0) = a$. We have begun to derive algebraic results from the geometric situation.

EXERCISES

1. Prove Proposition 1.

2. Prove Proposition 2.

3. Prove Proposition 3.

4. From the results available in this section, and from what we have previously proved about affine planes, is it possible to determine the equation of the line passing through the points (a, b) and (c, d)? If so, clearly outline the procedure by which this would be accomplished. If not, indicate what further information would have to be available before the equation could be found.

5. Actually carry out a coordinatization of the four-point affine plane. Here K would have 0 and 1 as its sole elements.

6. Carry out a coordinatization of an affine plane of nine points. Use $K = \{0, 1, 2\}$. If K is assumed to have the algebraic structure of the integers modulo 3 (that is, the field of three elements), determine if lines of the plane not parallel to the y-axis have an equation of the form $y = mx + q$, where m is their slope and $(0, q)$, the y-intercept. Carry out the coordinatization for various choices of the unit line, and x- and y-axis, and the origin. Do the coordinatizations obtained seem to be significantly different for any two sets of choices? Explain carefully.

4.2 MORE ALGEBRA OF THE AFFINE PLANE

We now have an affine plane A which has been coordinatized using a set K as described in the previous section; we have also defined a ternary operation F on K. We now define addition on K as follows:

(2)   For any elements a and b of K, set $a + b = F(a, 1, b)$.

It follows at once from (2) that any line with slope 1 and y-intercept $(0, q)$ has the equation $y = x + q$. As might be expected, addition in K has a geometric interpretation in A.

We associate the elements a and b of K with the points $P = (a, a)$ and $Q = (b, b)$, respectively, of the unit line in A. The point $(a + b, a + b)$ of the unit line associated with $a + b$ is found geometrically as follows: The line with slope 0 which contains $Q = (b, b)$ intersects the y-axis in the point $R = (0, b)$ (Fig. 4.4). The line with slope 1 which contains $(0, b)$ (that is, the line whose equation is $y = x + b$) intersects the line with equation $x = a$ (that is, the line which contains P and is parallel to the y-axis) in the point

$$U = (a, a + b).$$

The line with slope 0 which contains U intersects the unit line in

$$V = (a + b, a + b).$$

We could define $P + Q = V$ in a purely geometric way without reference to coordinates, using only three parallel classes of lines (the reader should determine how this is done and what three parallel classes are involved). If we set $P + Q = V$, with addition of points on the unit line defined according to the procedure outlined above, then the following facts can be proved geometrically.

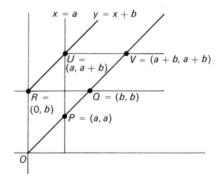

Fig. 4.4

Proposition 4

a) *If any two of the points P, Q, and V of the unit line are given, then there is one and only one point of the unit line for which* $P + Q = V$.

b) *If P is any point of the unit line, then* $P + 0 = 0 + P = P$.

It follows then that $K, +$ has all of the properties of a commutative group, except possibly for associativity and commutativity.

It appears from Fig. 4.4 that there is a similarity between the addition of P and Q, as defined earlier, and the addition of P and Q considered as "vectors," or "directed line segments." (This similarity is not surprising, since the diagram has been drawn to parallel the Euclidean situation.) We have no such things as line segments in the affine plane A, yet it is easily verified that the figures $RQVU$ and $RUOP$ are parallelograms at least in the sense that their opposite sides are parallel. If we assumed the usual properties of parallelograms in Euclidean geometry, then we would have $\overline{OP} = \overline{RU} = \overline{QV}$. Adding segments geometrically in the usual fashion, we see that $\overline{OQ} + \overline{OP} = \overline{OQ} + \overline{QV} = \overline{OV}$, which agrees with our earlier conclusion that V represents $Q + P$.

Having defined addition on K, found a geometric interpretation for addition, and related addition to the more familiar addition of segments, we turn to multiplication in K.

(3)   For any elements a and b of K, define $ab = F(a, b, 0)$.

It follows at once from (3) that the equation of the line with slope m which contains the origin is $y = mx$. Multiplication in K can be interpreted geometrically as follows: Let $P = (a, a)$ and $Q = (b, b)$ (Fig. 4.5). The line of slope 0 which contains Q intersects the slope line $(x = 1)$ in a point $U = (1, b)$. Then the line OU (with equation $y = xb$) intersects the line with equation $x = a$ in the point $R = (a, ab)$. The line of slope 0 which contains $R(y = ab)$ intersects the unit line in $V = (ab, ab)$.

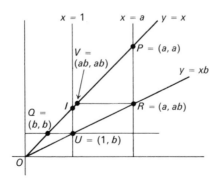

Fig. 4.5

Setting $P \cdot Q = V$, we can prove the following facts about multiplication of points on the unit line (and thus obtain corresponding results about multiplication in K).

Proposition 5

a) *Given any two of the points P, Q, and V of the unit line, the third point is uniquely determined by the equation $P \cdot Q = V$ provided that neither P nor Q is 0 while V is not 0.*

b) *For any point P of the unit line $P \cdot 0 = 0 \cdot P = 0$ and $I \cdot P = P \cdot I = P$ (recall that I is the unit point $(1, 1)$).*

Once again we can obtain a familiar equation from the geometric situation if we assume that the components of Fig. 4.5 have properties that they would have in the Euclidean situation. Specifically, $\triangle POR$ is similar to $\triangle IOU$ and $\triangle VOR$ is similar to $\triangle QOU$. This leads to the equalities

$$\overline{OP}/\overline{OI} = \overline{OR}/\overline{OU} = \overline{OV}/\overline{OQ}; \quad \text{hence } \overline{OP} \cdot \overline{OQ} = \overline{OI} \cdot \overline{OV},$$

where the ratios are of the lengths of the respective segments. If $\overline{OI}$ is assumed to have unit length, then we have $\overline{OP} \cdot \overline{OQ} = \overline{OV}$.

A vector in the coordinate plane R^2 is generally thought of as representable by a directed line segment; two directed line segments represent the same vector if they have the same length, point in the same direction, and are parallel. Put more formally, a vector can be considered as an equivalence class of directed line segments, where two directed line segments are equivalent, if they fulfill those conditions given for their representing the same vector.

A directed line segment d in R^2 can be designated by an ordered pair (P, Q) of points of R^2 where d begins at P and ends at Q. Two directed line segments (P, Q) and (P', Q') will then represent the same vector if PP' is parallel to QQ' and PQ is parallel to $P'Q'$ (Fig. 4.6).

Once again following the lead of the situation in R^2, we define a *directed line segment* in the affine plane A to be an ordered pair (P, Q) of points of A. We say that two directed line segments (P, Q) and (P', Q') in A are *equivalent* if PP' is parallel to QQ' and PQ is parallel to $P'Q'$. We also define any directed line segment to be equivalent to itself, and any "degenerate" line segment (P, P) to be equivalent only to all other degenerate

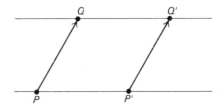

Fig. 4.6

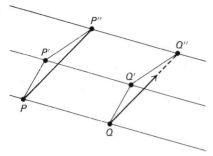

Fig. 4.7

line segments. We will use the more familiar notation $\overline{PQ}$ to designate the directed line segment (P, Q).

We would of course like to say that "is equivalent to" defines an equivalence relation on the set of all directed line segments of A; we could then define a *vector* to be an equivalence class of directed line segments. We have no trouble proving that any directed line segment is equivalent to itself (we have defined this to be true); clearly too, if $\overline{PQ}$ is equivalent to $\overline{P'Q'}$, then $\overline{P'Q'}$ is equivalent to $\overline{PQ}$. The difficulty arises when we try to show that $\overline{PQ}$ equivalent to $\overline{P'Q'}$ and $\overline{P'Q'}$ equivalent to $\overline{P''Q''}$ imply that $\overline{PQ}$ is equivalent to $\overline{P''Q''}$; $\overline{PQ}$ equivalent to $\overline{P'Q'}$ and $\overline{P'Q'}$ equivalent to $\overline{P''Q''}$ lead to the situation depicted in Fig. 4.7. Now it is clear from what is given that $\overline{PQ}$ is parallel to $\overline{P''Q''}$, but in order to be certain that $\overline{QQ''}$ is parallel to PP'' (and hence $\overline{PQ}$ is equivalent to $\overline{P''Q''}$), we must have some form of Desargues' Theorem. For we have that triangles $QQ'Q''$ and $PP'P''$ are perspective from a point (since PQ, $P'Q'$, and $P''Q''$ are parallel) with QQ' parallel to PP' and $P'P''$ parallel to $Q'Q''$. If we had Desargues' Theorem, then we could conclude that QQ'' is parallel to PP''. We can therefore say:

Proposition 6 *If A is Desarguesian, then "is equivalent to" defines an equivalence relation on the set of directed line segments of A.*

We will assume henceforth in this section that A is Desarguesian. We define an equivalence class of directed line segments of A to be a *vector* of A. We denote the equivalence class of $\overline{PQ}$ by **PQ**.

We will shortly define an addition for vectors. Before doing so, we prove a lemma.

Lemma *Given any point R and any vector **PQ**, there is a point U such that **RU** = **PQ**.*

Proof We distinguish two cases.

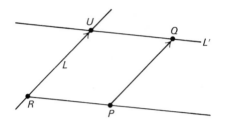

Fig. 4.8

CASE 1 *R*, *P*, and *Q* are noncollinear (Fig. 4.8). Let *L* be the line which contains *R* and is parallel to *PQ*, and *L′* be the line which contains *Q* and is parallel to *PR*. Set *U* equal to the point of intersection of *L* and *L′*.

CASE 2 *R*, *P*, and *Q* are collinear. Let *L* be any line which contains *R*, other than *RP*, and let *L′* and *L″* be the lines which contain *P* and *Q*, respectively, and are parallel to *L*. Let *P′* be any point of *L′*, and *H* be the line which contains *P′* and is parallel to *RP*. Then *H* and *L″* intersect in some point *Q′* and **PQ** = **P′Q′**. Apply Case 1 to find *U* so that **RU** = **P′Q′**. Then **RU** = **PQ**.

We define addition for vectors as follows: Given **PQ** and **P′Q′**, find *U* so that **QU** = **P′Q′**. Set

$$\mathbf{PQ} + \mathbf{P'Q'} = \mathbf{PU} \quad \text{(Fig. 4.9)}.$$

We leave the proof of the following proposition to the reader.

Proposition 7 *The set of vectors with addition defined as above is a commutative group.*

The following proposition gives another important algebraic consequence of Desargues' Theorem (which we have assumed for *A*).

Proposition 8 *Given any three elements a, b, and c of K,*

(4) $$F(a, b, c) = ab + c;$$

hence the equation of a line with slope m and y-intercept (0, q) is y = mx + q.

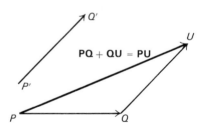

Fig. 4.9

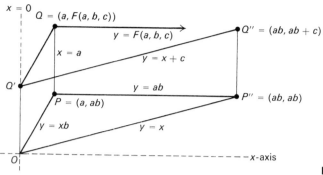

Fig. 4.10

Proof If a, b, or c is 0, or $b = 1$, then (4) is true by equalities previously established. We therefore assume that a, b, and c are not 0 and $b \neq 1$. Consider Fig. 4.10. Clearly (4) will hold if and only if the line whose equation is $y = F(a, b, c)$ contains the point $(ab, ab + c)$; this latter condition will hold if and only if PP'' and QQ'' are parallel. But PP'' is parallel to QQ'' by Desargues' Theorem applied to triangles $PO'P''$ and $QQ'Q''$.

Proposition 9 *For any elements a, b, and c of K,*

a) $(b + c)a = ba + ca$, *and*
b) $a(b + c) = ab + ac$.

That is, multiplication is distributive over addition in K.

Proof We prove (a) using Fig. 4.11. We want to show that the point $U'' = (a + b, (a + b)c)$ coincides with $U' = (a + b, ac + bc)$. Since (a)

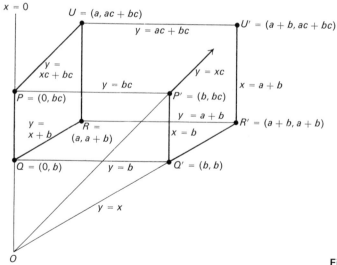

Fig. 4.11

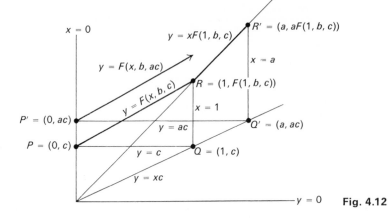

Fig. 4.12

holds trivially if a, b, or c is 0, or if $c = 1$, we assume that a, b, and c are not 0 and $c \neq 1$. Using vectors, we obtain $\mathbf{P'U'} = \mathbf{P'Q'} + \mathbf{Q'R'} + \mathbf{R'U'}$. Now $\mathbf{P'Q'} = \mathbf{PQ}$, $\mathbf{Q'R'} = \mathbf{QR}$, and $\mathbf{R'U'} = \mathbf{RU}$; hence $\mathbf{P'U'} = \mathbf{PQ} + \mathbf{QR} + \mathbf{RU} = \mathbf{PU}$. Then slope $PU = $ slope $P'U' = c$. But slope $OP' = c$; hence O, P', and U' are collinear. Therefore OP' intersects $R'U'$ in U'; hence $U'' = U'$. This completes the proof of (a).

Figure 4.12 is used to prove (b). We leave the details of the proof as an exercise.

Although we have proved some of the algebraic consequences of the axioms for an affine plane and of Desargues' Theorem, it requires considerably more work to arrive at the following fundamental theorem.

Proposition 10 *The affine plane A is Desarguesian if and only if the set K, with addition and multiplication as defined earlier in this section, is a (not necessarily commutative) field.*

Both because of limitations of space, and because we do not want to presume too much algebraic background on the part of the reader, we omit any further presentation of the proof of Proposition 10. For a more complete discussion of the algebraic consequences of Desargues' Theorem, as well as a good bibliography of literature available on the topics considered in this chapter, the reader might see the paper which served as a model for much of the presentation of this chapter: R. H. Bruck's, "Recent Advances in the Foundations of Euclidean Geometry."*

---

* R. H. Bruck, "Recent Advances in the Foundations of Euclidean Geometry," *American Mathematical Monthly*, **62** (No. 7), 2–17 (available as Number 4 of the Slaught Memorial Papers from the Mathematical Association of America).

The following are corollaries of Proposition 10.

Corollary 1 *Any Desarguesian affine plane is isomorphic to an affine plane $A(F)$ over some (not necessarily commutative) field F.*

Corollary 2 *If A is a Desarguesian affine plane, then $\pi(A)$ is also Desarguesian.*

Proof If A is Desarguesian, then A is isomorphic to $A(F)$ for some field F. But then $\pi(A)$ is isomorphic to $\pi(F^3)$, which is Desarguesian.

We conclude this section with a classical example of an affine plane which is not Desarguesian.

Example 2 Let R^2 be the coordinate plane, but with lines modified as follows: Any regular line of R^2 with nonpositive slope, or with slope undefined, will be admitted as a line without modification. A line with equation $y = mx + q$, where m is positive, will be modified to the set consisting of the points (x, y) with

$$y = mx + q, \qquad y < 0,$$

and

$$y = (m/2)x + q, \qquad y \geq 0.$$

Thus, lines of positive slope are refracted as they cross the x-axis (Fig. 4.13). We leave it to the reader to show that R^2, with the linear structure thus defined, is an affine plane. Figure 4.14 gives a configuration in which two triangles are perspective from a point, but are not perspective from a line; hence Desargues' Theorem does not apply to this affine plane.

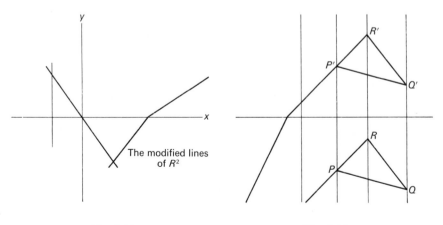

Fig. 4.13   Fig. 4.14

EXERCISES

1. What three parallel classes of lines are involved in defining $P + Q$ for points P and Q of the unit line? Define $P + Q$ geometrically without reference to the coordinates which have been assigned to the points of A.

2. Prove Proposition 4. (Figure 4.4 should be of help in the proof of both (a) and (b).)

3. Prove Proposition 7. Specifically, prove each of the following:
   a) Prove that addition of vectors is well defined; that is, prove that $\mathbf{PQ} = \mathbf{P'Q'}$, and $\mathbf{RU} = \mathbf{R'U'}$ implies $\mathbf{PQ} + \mathbf{RU} = \mathbf{P'Q'} + \mathbf{R'U'}$.
   b) Prove that addition of vectors is commutative; that is, $\mathbf{PQ} + \mathbf{RU} = \mathbf{RU} + \mathbf{PQ}$.
   c) Prove that addition of vectors is associative.
   d) Prove that $\mathbf{PP}$ acts as a zero vector and $\mathbf{QP} = -\mathbf{PQ}$.

4. Prove (b) of Proposition 9 (use Fig. 4.12).

5. Consider Example 4.
   a) Prove that R^2 with the linear structure described in this example is an affine plane.
   b) Since this affine space is not Desarguesian, it does not have a complete set of dilatations. Try to find three distinct collinear points C, P, and Q, for which there is no homothety with center C that maps P onto Q, or two points P' and Q', for which there is no translation that maps P' onto Q'.

6. Let A be the Desarguesian affine plane coordinatized in this chapter, using the set K. Given points (a, b) and (c, d) of A, what is the equation of the line which contains these points?

7. The condition needed in order for "is equivalent to" to define an equivalence relation on the set of all vectors of A is actually weaker than Desargues' Theorem. Try to state this condition precisely, and indicate in what respect it differs from Desargues' Theorem.

**Part II Euclidean Geometry
from the
Synthetic Viewpoint**

5 Introduction.
The Axioms
of Order

It is only a slight exaggeration to say that mathematics is an attempt to put together a jigsaw puzzle without knowing exactly what the puzzle is supposed to look like when it is finished. As mathematics has developed historically, little bits and pieces of information have been gathered and used independently, until some man, or men, could unify all or some of the diverse pieces by showing that they all shared some unifying aspect, that is, that they could all be brought under one roof by viewing them properly. Thus, if each bit of knowledge is viewed as a piece of a jigsaw puzzle, we could say that pieces of the puzzle were gathered one by one until someone was able to say, "These different pieces would make sense if we viewed them all as part of a picture of a sailboat." Then, knowing better what we are looking for, we can more readily find and fit into place other pieces of the puzzle.

Imperfect knowledge, of course, is subject to many interpretations. For example, while one person may see a sailboat in the many different pieces, another may see a landscape. Which person would be right? Perhaps both. Indeed, it has often happened that viewing the same information from different viewpoints has led to different and equally rewarding studies, not only in mathematics, but in many other disciplines.

Nevertheless, it is not altogether true that mathematicians studying seemingly unrelated bits of knowledge have no idea what the jigsaw puzzle should be like. The fact is that most, if not all, mathematics has ultimately had its inspiration in the real world. Much of mathematics is intended to embody, or express, what we see in the world around us; hence, by studying the world, we often glimpse what the jigsaw should be like. Many results of Euclidean geometry were known before Euclid. The Egyptians, for example, used a special case (the 3-4-5-right triangle) of the Pythagorean Theorem to obtain right angles perhaps 2400 years before Pythagoras, who lived about 600 B.C., and 2700 years before Euclid. Parallel lines and other interesting geometric configurations are found in Egyptian art dating back to 4000 B.C. Certainly, then, much of what is included in Euclidean geometry was known, at least in embryonic form, thousands of years before Euclid compiled his *Elements*. Yet, even though many pieces of the geometric jigsaw puzzle were not available in a unified form before Euclid, it cannot really be said that Euclid, or even the ancient Egyptians, had no idea of the whole puzzle, for they felt that their geometry was the geometry of the real world.

The fact that Euclid and, perhaps more important, many after Euclid ascribed reality to Euclidean geometry had a profound effect on the development of geometry. We shall discuss this point at greater length after learning something about Euclid himself.

Almost nothing about Euclid's life is certain. One known fact, however, is that Euclid founded a school in Alexandria about 300 B.C. While Euclid was apparently a competent and creative mathematician in his own right— he published a number of other works besides the *Elements*—his great contribution to mathematics rests in his magnificent compilation, the *Elements*, of work that had been done primarily by others. There were other *Elements*—that is, compilations of geometric knowledge—before Euclid, the first being that of Hippocrates of Chios in the fifth century B.C.; but Euclid's *Elements* quickly replaced this and all others.

Euclid tried to make his *Elements* a strictly logical derivation, based on axioms, definitions, and postulates, of much of the mathematical theory of his time*. Although Euclid compiled the work of his colleagues and predecessors, much of the brilliance of the *Elements* must be credited directly to him. The arrangement of the books and the material within the books to form a coordinated and smoothly flowing whole is one of Euclid's great contributions; at times, he found new proofs to theorems already known in order to fit these theorems more naturally into the whole.

The first four books of the *Elements* deal with plane geometry, the same plane geometry which we shall soon investigate, and, indeed, which the reader has encountered much earlier in his schooling. The fifth book deals with one of the most important discoveries of Euclid's day, Eudoxus's

---

* For Euclid there was a technical distinction between *axiom* and *postulate*. An axiom embodied principles which Euclid felt no one could deny without being illogical; that is, an axiom is intrinsically true and applicable to any mathematical system. A postulate, on the other hand, is used to define a *particular* mathematical system.

theory of proportions, while Book Six applies this theory to triangles. Today the results of Book Six would be studied as results pertaining to similar triangles. Books Seven to Nine deal primarily with number theory, Book Ten with a geometrical classification of quadratic roots, while Books Eleven through Thirteen deal with solid geometry.

In reality, Euclid did not achieve a development based solely on his axioms and postulates. Analyses of his work have shown that he assumed several postulates which he left unstated. Moreover, certain of Euclid's definitions are meaningless and unnecessary. Euclid, for example, tried to define the terms *line* and *point*, which were actually primitive terms in his axiom system; that is, they could not be defined, but rather should have been used just to formulate the axioms and later definitions. Nevertheless, since Euclid felt that *line* and *point* referred to real objects, it is not surprising that he tried to express in his definitions what these real objects were.

Even though Euclid did not possess the mathematical sophistication of the twentieth century, his *Elements* stood for over 2000 years as the model of mathematical excellence. The *Elements* can even today be studied to advantage; it is fair to say that they are the most influential mathematical work, and, in fact, one of the most influential pieces of literature, ever written.

One of Euclid's original contributions to the *Elements* and to mathematics in general is his celebrated Fifth Postulate:

If two straight lines in a plane meet another straight line in the plane so that the sum of the interior angles on the same side of the latter straight line is less than two right angles, then the two straight lines will meet on that side of the latter straight line.

Euclid uses this postulate to prove such fundamental results as:

*From a point outside a line, one and only one line can be constructed parallel to the given line;**

and

The sum of the interior angles of a triangle is equal to two right angles.

Although there is evidence to suggest that Euclid was not altogether satisfied with his Fifth Postulate, he was, in fact, the first mathematician to realize that some postulate was necessary in order to have a theory of parallels, and that the existence of parallels did not flow from the other assumptions of geometry. The Fifth Postulate was one of the most distinctive, and soon one of the most controversial, features of Euclidean geometry.

Even in Euclid's time there was a fair amount of discussion of the Fifth Postulate. Euclid himself did not seem satisfied with it, and it is certainly

---

* Although many speak of this as Euclid's Fifth Postulate, it is actually a theorem.

not as appealing or self-evident as most of his other axioms or postulates. Though many geometers tried to find a more pleasing form of the Fifth Postulate, and others tried to prove it from the other Euclidean assumptions, no one succeeded.

After the decline of Greek civilization, not much was done about the Fifth Postulate until near the end of the seventeenth century. At that time, the English mathematician John Wallis (1616–1703) claimed to have proved Euclid's postulate, but his proof depends on an assumption equivalent to Euclid's. The Italian Jesuit priest, Girolamo Saccheri, tried, in about 1733, to prove Euclid's postulate by showing that postulates contrary to it led to contradictions. The contradiction that Saccheri arrived at was the statement that the sum of the interior angles of a triangle is less than two right angles, a perfectly valid theorem in non-Euclidean geometry. Saccheri, in fact, developed an extensive body of theorems in non-Euclidean geometry and, had he been a bit more open-minded, he might have been remembered as the father of this study.

To understand the close-minded attitude of Saccheri and Wallis, and many of the contemporaries of Euclid, we must recall that the specialization of knowledge as we know it today is a comparatively recent phenomenon. Nowadays, not only can one study mathematics to the exclusion of almost every other branch of knowledge, but one can also often specialize within mathematics itself to the virtual exclusion of other branches of mathematics. This observation is, of course, true in most fields of study. In Euclid's time, however, and continuing through the Renaissance, educated men were supposedly masters of many disciplines; moreover, the disciplines themselves intertwined a great deal. Mathematics in ancient Greece, as elsewhere in the ancient world, often had mystical and philosophical overtones. In the middle ages we find theologians and philosophers discoursing on the sciences, basing what they say on philosophical rather than empirical considerations; in this, they were following the lead of Aristotle and other eminent Greeks. However, inasmuch as mathematics and science were supposed to describe reality, and philosophy and theology were the most real disciplines of all, one can easily understand the subordination of all other branches of knowledge to theology or metaphysics.

The brilliant theory of Eudoxus was largely a response to the paradoxes raised by Zeno of Elea (ca. 450 B.C.) which, in turn, were philosophical in origin, coming from arguments concerning the reality of change. One of Zeno's most famous paradoxes is the following:

Suppose I want to go from a point A to a point B along a straight line. Then I must first travel half the distance from A to B. If B_1 is the midpoint between A and B, then I must also travel half the distance from A to B_1. But this process never ends; therefore my traveling from A to B can never begin.

In the books on solid geometry (eleven through thirteen) Euclid presents the work of Theaetetus on the five regular solids which occupy a central place in the cosmology of Plato.

Of course Plato's cosmology was not of more than classical interest to the philosophers of the Middle Ages, and the paradoxes of Zeno were resolved to the satisfaction of most; yet, so closely did many people consider Euclidean geometry to be tied to the real order of things that no one dared question its intrinsic correctness and suggest, for example, that the geometry of the universe might be better expressed by a set of axioms other than Euclid's. In particular, the parallel postulate was held in such high esteem that the great mathematician Carl Friedrich Gauss (1777–1855) was afraid to publish, for fear of the uproar that it would have caused, his discovery that there are perfectly decent non-Euclidean geometries. It was not until the nineteenth century that non-Euclidean geometries were systematically investigated.

The latter part of the nineteenth century and the twentieth century have seen a strong revitalization and reinvestigation of the axiomatic method in mathematics. The very foundations of mathematics itself, as well as the foundations of its various branches, have undergone extensive study. Of particular concern to us is the fact that a number of mathematicians tried to find a rigorous axiomatization of Euclidean plane geometry since, as we have seen, such was not achieved by Euclid. Among the many who worked on this problem were the Germans Pasch, Schur, and Hilbert; the Italians Peano and Pieri; and the Americans E. H. Moore and Oswald Veblen. This list is by no means exhaustive, and, indeed, there are still today many mathematicians exploring the foundations of geometry. In the next section we shall investigate some of the methods employed in axiomatizing Euclidean plane geometry, and we shall select a particular axiomatization to explore at greater length.

EXERCISES

1. This section has been a necessarily brief history of Euclidean geometry. There are, of course, many points that could have been made, and names that could have been mentioned, that were not. Prepare a short report on each of the following. Be sure to point out the pertinence of each topic to the study of Euclidean geometry.

a) Hippocrates of Chios
b) Pythagoras
c) Eudoxus
d) Quadrature of the circle
e) Georg Riemann
f) Eugenio Beltrami
g) Felix Klein
h) Giuseppe Peano
i) Johann Bolyai
j) Nikolai Lobachevski

2. We showed in Part I that any Desarguesian affine plane is isomorphic to an affine plane over some field F. In order to have Euclidean plane geometry, the field must be the field of real numbers. Show that in $A(F)$, the affine plane on the field F, the unit line (that is, the line with equation $y = x$) is isomorphic to F in a meaningful way. Since any two lines of an affine plane are isomorphic (why?), each line of $A(F)$ must in some sense look like F. Try to find additional hypotheses which might make a Desarguesian affine plane isomorphic to the affine plane on the field of real numbers. Your hypotheses should be directed toward making lines look like the real number line. What are some set-theoretic and geometric properties of the real number line?

3. Read one or more of the following:

   a) J. W. Young, *Lectures on the Fundamental Concepts of Algebra and Geometry*, New York: Macmillan, 1961, Lectures I through V.

   b) M. C. Gemignani, "On the Geometry of Euclid," *The Mathematics Teacher*, **60** (February 1967), 160–164.

   c) H. Von Helmholtz, "On the Origin and Significance of Geometrical Axioms," *The World of Mathematics*, Vol. 1, J. Newman (ed.), New York: Simon and Schuster, 647–666.

5.2 NEW APPROACHES TO AN OLD SUBJECT

As we have already seen, Euclid did not succeed in adequately axiomatizing his geometry. The problem of axiomatizing what is now known as *Euclidean plane metric geometry* has been solved by a number of men using a variety of methods.

How can axiom systems differ if they both describe the same thing? First, an undefined term in one axiomatization may be definable relative to another axiomatization. Secondly, an axiom of one system may be a theorem in another. Having two distinct axiomatizations of the same thing is analogous to the following: Suppose a certain town has but one house with a pine tree on the front lawn, and this house is located at 30 Main Street. Then someone inquiring how to locate the house could be told that it is the only house in town with a pine tree on the front lawn, or he could be given the explicit address. Either bit of information would be sufficient to determine the house precisely, and the information that was not given could be learned once the house was found. From a practical point of view, knowing that the house is at 30 Main Street may be better than knowing that the house has a pine tree in front, since, as a rule, we could find the house much faster with the former information. However, it may be that in certain instances the information that the house is the only house with a pine tree in front will be more valuable than knowing its address. Similarly, two different axiomatizations of a mathematical system may each serve some important purpose.

How do we know whether or not two systems of axioms define the same thing? Essentially, if D and D' are two systems of axioms and D is a theorem (or set of theorems) in the system defined by D', and D', in turn, is a theorem in the system defined by D, then D and D' define the same mathematical system. In such a case, anything that can be proved from D can be proved from D' as well, and, conversely, anything that can be proved from D' can also be proved from D. It is not necessarily true, however, that two axiomatizations of the same system have the same number of undefined terms, or even the same number of axioms. The axiomatization of Euclidean solid geometry devised by the Italian Mario Pieri and published in 1899 uses only one undefined term and seventeen axioms, while David Hilbert uses four undefined terms and twenty-one axioms to do the same thing.

Which, we may ask, is better, Hilbert's set of axioms, or Pieri's? We would then have to answer the question, "Better for what?" From the viewpoint of the number of undefined terms and the number of axioms, we might say that Pieri's axiomatization is better than Hilbert's. On the other hand, from the viewpoint of giving insight into the nature of Euclidean geometry and having intuitive appeal, Hilbert's axiomatization is far better. Which is more elegant, beautiful, esthetically pleasing, etc. is largely a matter of opinion. In reality, the fundamental methods of these two mathematicians in approaching Euclidean geometry are very different. It might even be said that they represent opposite extremes. This point will become clearer after we see which approaches to axiomatizing Euclidean geometry are possible.

Geometries can be characterized algebraically by means of groups of related collineations or *transformations*. The geometry itself is concerned with properties which are left unchanged by any collineation in the characterizing group. For example, Euclidean geometry studies properties which are left unchanged by so-called *rigid motions*. A rigid motion, intuitively speaking, occurs when an object is moved without altering its shape. This idea of rigid motion is often expressed as a *principle of superposition*; that is, two objects are equivalent from the viewpoint of Euclidean geometry, or are *congruent*, if one can be rigidly moved in such a manner as to be exactly superimposed on the other object. Such a principle of superposition is, of course, highly informal, and, unfortunately, as it is sometimes bandied about, has virtually no mathematical content; but it does give a crude glimpse of the type of transformation with which Euclidean geometry is concerned. More formally, a *rigid motion* is a collineation which preserves either distance between points or congruence of pairs of points, with either the distance between points or the notion of congruence being added axiomatically. (Not all collineations of the usual coordinate plane R^2 are rigid motions. For example, the dilatation defined by $(x, y) \rightarrow (2x, 2y)$ does not preserve distances in R^2. Any rigid motion of R^2 is a collineation of the affine geometry of R^2, but not every collineation of the affine geometry is a rigid motion. Euclidean geometry is, as the reader already knows, a special case of affine geometry.) Examples of properties preserved by rigid motions are length, area, and angle measurement.

If a geometry has been characterized in some manner other than specifying its special group of transformations, we can then find the collineation group which characterizes the geometry. But since some transformation group does characterize any geometry, we could also start with the transformation group, and then find other methods of specifying the geometry.

The approach to a geometry through its special group of transformations is, of course, essentially algebraic in nature. Not all groups are suitable for defining a geometry. If some group is sufficiently "nice," then we can define a geometry for which that group is the group of transformations. And, in fact, if the group has just the right properties, the geometry will then be Euclidean solid metric geometry. While the concept *motion* in Pieri's system of axioms for solid geometry is left undefined (it is the only primitive term), his axioms are intended to embody those properties which characterize the rigid motions of Euclidean geometry. Pieri would thus approach the group of transformations first and would then pass to the geometry.

How might one obtain a line from a group of transformations? Pieri begins with a set of *motions* on a set S (essentially the motions are functions from S into S). The elements of S are called *points*. Pieri's Fifth Axiom states:

If there is a motion which leaves three points A, B, and C fixed, then every motion which leaves A and B fixed leaves C fixed.

This axiom can be expressed in terms of functions as follows:

*If there is a motion which has A, B, and C as fixed points, then C is a fixed point for **any** motion which has A and B as fixed points.*

The *line AB* can then be defined as the set of all points which are fixed points of all motions which have A and B as fixed points.

Another of Pieri's axioms is as follows:

Axiom 10 *If A, B, and C are three points which are not collinear, then there is a motion which has A and B as fixed points and which maps C onto another point of the plane determined by A, B, and C. (**Plane**, like **line**, is a term defined in terms of **motion**.)*

Informally, this axiom expresses the idea that if A, B, and C are points in the same plane, then there is a *symmetry* of that plane with respect to the line AB; that is, the plane can be rotated about the line AB, keeping the line AB fixed. We might illustrate Axiom 10 by means of a picture such as that given in Fig. 5.1.

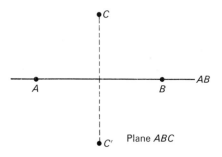

Plane *ABC*

Fig. 5.1

Hilbert's and Veblen's approaches, on the other hand, characterize the geometry first. Once the geometry is defined, the group of transformations can be found. While Pieri's term *motion* can be considered to be algebraic in nature, Hilbert's undefined terms, *point, line, between,* and *congruent,* are distinctly geometric in flavor. Hilbert's axioms give properties which one would find geometrically obvious. For example, the following are two of Hilbert's axioms:

Axiom I.6 *If two points A and B of a line L are in a plane, then every point of L is in that plane.*

Axiom II.4 *Let A, B, and C be three noncollinear points and L be a line in the plane ABC which does not contain A, B, or C. Then if L contains a point of the segment AB* (segments are defined by Hilbert using the Axioms of Order), *L will also contain a point of the segment BC or a point of the segment AC* (Fig. 5.2).

Almost all of Hilbert's axioms are as geometrically appealing as the two cited.

In addition to the extremes of Pieri and Hilbert, there are some axiomatizations which use a certain number of algebraic axioms together with geometric axioms. Certain topological concepts, such as *separation* properties and *completeness,* have been brought into some more recent axiomatizations of geometry, but we are not prepared to discuss this here.

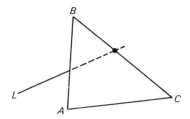

Fig. 5.2

The set of axioms we will use for Euclidean plane geometry was the work of the American Oswald Veblen. This axiomatization first appeared in the *Transactions of the American Mathematical Society* in 1904. Veblen's approach, like Hilbert's, is very clearly geometric in character (as opposed to algebraic or topological). In the form in which we will present these axioms, there are three undefined concepts: *between*, *point*, and *congruent*. There are fourteen axioms used to define plane geometry, and we would need only two more axioms to define Euclidean solid geometry.

The manner and order in which we will present Veblen's work is basically that of Veblen's paper, "The Foundations of Geometry," which appears in the Dover publication *Monographs on Modern Mathematics,* 1955. We suggest that the entire *Monographs* might be a valuable addition to the reader's mathematical library since the work contains excellent material on many of the classical problems in mathematics.

EXERCISES

1. Given below are certain axioms of David Hilbert. Indicate what geometric fact each is intended to convey. Draw an appropriate figure to illustrate the axiom.

   a) Let $\overline{AB}$ and $\overline{BC}$ be two segments on a line L such that $\overline{AB}$ and $\overline{BC}$ share only the point B in common. Furthermore, let $\overline{A'B'}$ and $\overline{B'C'}$ be segments on a line L' such that $\overline{A'B'}$ and $\overline{B'C'}$ share only B' in common. Then if $\overline{AB}$ is congruent to $\overline{A'B'}$ and $\overline{BC}$ is congruent to $\overline{B'C'}$, we have $\overline{AC}$ congruent to $\overline{A'C'}$.

   b) Let A_1 be any point on the line AB which lies between A and B. Take the points $A_2, A_3, A_4, \ldots$, so that A_1 lies between A and A_2, A_2 lies between A_1 and A_3, A_3 lies between A_2 and A_4, etc. Moreover, let the segments $\overline{AA_1}$, $\overline{A_1A_2}$, $\overline{A_2A_3}$, etc., be congruent to one another. Then, among the points $A_2, A_3, A_4, \ldots$, there is always a point A_n such that B lies between A and A_n.

   Follow the same directions as above for the following axioms of Pieri. Express each of these axioms by using the function terminology.

   c) If A and B are distinct points, then there is a motion which leaves A fixed and transforms B into another point of the line AB.

   d) If A and B are distinct points, then there is a motion which transforms A into B and which leaves fixed a point of the line AB.

2. Discuss the following question and try to arrive at a satisfactory answer: A set of axioms is supposed to define a mathematical system. How do we know when we have enough axioms? For example, how does a mathematician trying to axiomatize Euclidean plane geometry know when his axioms have done the job? This, in turn, leads to another question: Do mathematical systems exist in some ideal form apart from the mathematician, and is the mathematician merely trying, by means of his axioms, to capture that external reality? The

reader should not expect a cut-and-dried answer to these questions, questions which are not at all simple, even for the expert mathematician, and which have inspired much, and sometimes heated, debate; but the reader should think about them.

3. We define the *distance* $D(P, P')$ between two points $P(a, b)$ and $P'(a', b')$ of the coordinate plane R^2 by

$$D(P, P') = \big((a - a')^2 + (b - b')^2\big)^{1/2}.$$

(This definition, as the reader may recall, is inspired by the Pythagorean Theorem.) Prove that the dilatation of R^2 defined by $(x, y) \to (2x, 2y)$ does not preserve distance between points; that is, $D(P, P') \neq D(Q, Q')$, where P and P' are mapped onto Q and Q', respectively. See Example 9 of Chapter 1 for the general form of a collineation of R^2. Try to find a necessary and sufficient condition for a collineation of R^2 to preserve distances.

4. Read one or more of the following:
   a) J. W. Young, *Lectures on the Fundamental Concepts of Algebra and Geometry*, Lectures XIII–XV.
   b) D. Hilbert, *The Foundations of Geometry*, trans. by E. Townsend, La Salle, Ill.: Open Court, 1947.
   c) R. Wilder, "The Role of the Axiomatic Method," *American Mathematical Monthly*, **74** (February 1967): 115–127.

5.3 THE AXIOMS OF ORDER

In Part I we began with the notion of a linear space and then added hypotheses to this basic notion to form more complex geometric structures. In the present study of Euclidean plane geometry, we will not use linear spaces as our point of departure. There are two reasons for this. First, by using Veblen's approach we will give the reader a view of another technique by which geometry may be approached. The contrast between the methods of Part I and Part II may give the reader more insight into how geometries are defined and studied than a straightforward continuation of the techniques of Part I. Second, as we tried to point out in Exercise 2 of Section 5.1, a good deal of care must be taken to make a line in an affine geometry look like the real number line, if we are to have Euclidean geometry. Veblen's approach gives a great deal of insight into the structure of lines in Euclidean geometry.

The same warning applies in this section, however, that applied in Part I. Let the reader beware of confusing his intuitive feelings with valid proof based only on the axioms and primitive terms. We will assume, as in Part I, that there is an underlying set of points on which the structure is being defined. Even though *point* will refer to an element of the underlying set, it is still an undefined term since the set is not specified. For the moment, the only other undefined term will be *between*. We now state our first axioms.

Fig. 5.3

Fig. 5.4

Axiom 1 *If a point B is between the points A and C, then A, B, and C are distinct.*

Axiom 2 *If a point B is between the points A and C, then C is not between B and A.*

If *B* is between *A* and *C*, we will write *A-B-C*. Thus, Axiom 2 states *A-B-C* implies we do not have *B-C-A*.

Axiom 1 is illustrated by Fig. 5.3. Observe that Axiom 2 is intended to prevent *lines* (Definition 1) from being circular, that is, from presenting an appearance like that shown in Fig. 5.4. We now present our first set of definitions.

Definition 1 *If A and B are distinct points, then the **line** AB is defined to consist of all points X such that A-X-B, X-A-B, A-B-X, X = A, or X = B* (Fig. 5.5).

*We define the **segment joining** A **and** B to be the set of all points X such that A-X-B. We denote this segment by (AB). The points A and B are said to be **end points** of (AB).* (Note, however, that *A* and *B* are not themselves points of (*AB*) since, by Axiom 1, *A-X-B* implies *X* is distinct from *A* and *B*.)

*The segment (AB), together with its end points, is called the **interval joining** A **and** B. We denote this interval by [AB].*

*Points contained in the same line are said to be **collinear**. Points not collinear are said to be **noncollinear**.*

With just two moderately weak axioms we are a long way from our goal of the Euclidean plane. We do not even have a linear space yet, since models can be found which satisfy Axioms 1 and 2 but within which two lines intersect in more than one point. The reader should find such a model.

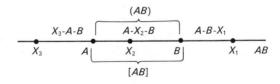

Fig. 5.5

From Definition 1 we can immediately conclude:

Proposition 1 *If A and B are distinct points, then AB consists of all points X such that X is a point of [AB], B is a point of (AX), or A is a point of (XB).*

We now add three more axioms.

Axiom 3 *If C and D are distinct points of the line AB, then A is contained in CD.*

Axiom 4 *If A and B are distinct points, then there is a point C such that A-B-C.*

Intuitively, Axiom 4 states that lines have no end points.

Example 1 Let *Z* be the set of integers. Let the integer *B* be between the integers *A* and *C* if $A < B < C$, or $C < B < A$. Then *Z* serves as a model which satisfies Axioms 1 through 4. Clearly we are still a long way from Euclidean geometry.

Axioms 1 through 4 are simply stated and are assented to easily. The following axiom is a bit more complicated but still embodies a rather obvious geometric property.

Axiom 5 *If A, B, and C are three noncollinear points, and D and E are points such that C is in (BD) and E is in (CA), then there is a line which contains D, E, and a point of (AB). (See Fig. 5.6.)*

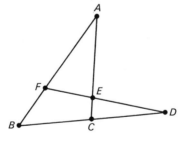

Fig. 5.6

The following example gives a model which satisfies Axioms 1–4 but not Axiom 5.

Example 2 Let *W* consist of all points of the coordinate plane both of whose coordinates are integers. We let *B* be between *A* and *C* if *B* is a point of the straight line segment (without end points) joining *A* and *C*. Consider the points presented in Fig. 5.7. Even though *W* satisfies Axioms 1–4 and the points satisfy the hypotheses of Axiom 5, the line (1, 0)(0, 1) fails to contain any points of $((-1, 0)(0, 2))$; thus the system does not satisfy Axiom 5.

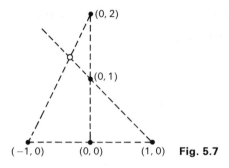

Fig. 5.7

In making use of Axiom 5, it would be easy to let our intuition outrun what we are logically entitled to say. For example, it appears that the line which contains *D*, *E*, and a point of (*AB*) is the line *DE*, but Axiom 5 does not explicitly say this. We have not yet proved that any two lines which contain *D* and *E* are necessarily the same. We will prove this fact soon, however.

In order to avoid situations such as that in Example 1 where there is only one line, we adopt the following axiom.

Axiom 6 *There are three noncollinear points.*

We will see that even with only Axioms 1–6 we can prove many of the properties we expect Euclidean geometry to have. The following example presents a system in which Axioms 1–6 hold, but which differs from the Euclidean plane in many important respects.

Example 3 Let *Y* consist of all points of the coordinate plane R^2, both of whose coordinates are rational numbers. Let *B* be between *A* and *C* if *B* is on the line segment joining *A* and *C*. We leave it to the reader to verify that Axioms 1 to 6 are satisfied by *Y*.

Since we have omitted a large set of points from R^2 to form *Y*, we would expect the geometric structure of *Y* to be quite different from that of the Euclidean plane. Such is, indeed, the case. For example, the lines of *Y* contain only countably many points, whereas the lines of the usual geometry on R^2 are uncountably infinite sets. Moreover, the lines of *Y* are not connected, that is, they have many gaps where points with at least one irrational coordinate belong.

We now begin to prove properties of the structure defined by Axioms 1 through 6.

Proposition 2 *Suppose A-B-C. Then we also have C-B-A, but none of the following: C-A-B, B-A-C, A-C-B, or B-C-A.*

Proof By Axiom 1, *A*, *B*, and *C* are distinct. Since *A-B-C*, *A* is contained in *BC*. Therefore by Axiom 3, *B* is contained in *CA*. By definition of *CA*, we must have *C-A-B*, *C-B-A*, or *B-C-A*. By Axiom 2, however, *A-B-C* excludes *B-C-A*. Moreover, if we have *C-A-B*, then again by Axiom 2, we could not have *A-B-C*. Therefore it follows we have *C-B-A*, but neither *B-C-A* nor *C-A-B*. Now if we had *A-C-B*, then from what we have already proved, we would have *B-C-A*, which has been excluded. Also, *B-A-C* implies *C-A-B*, which is impossible. This completes the proof.

Corollary 1 If *A* and *B* are distinct points, then $AB = BA$, $(AB) = (BA)$, and $[AB] = [BA]$.

Corollary 2 If *A-B-C*, then *A*, *B*, and *C* are all contained in the lines *AB*, *BC*, and *CA*.

The proofs of these corollaries are left as exercises.

Proposition 3 *Given two distinct points, there is one and only one line which contains them both.*

Proof Let *A* and *B* be distinct points. *A* and *B* are both points of $AB = BA$. Let *C* be any point of *AB* other than *A*. We now show $AB = AC$. Let *X* be any point of *AB*. If *X* is either *C* or *A*, then *X* is a point of *AC*; assume *X* is neither *C* nor *A*. Since *C* is in *AB*, we have by Axiom 3 that *A* is in *CX*. Consequently, *C-A-X*, *A-C-X*, or *C-X-A*. Whichever of these cases holds, it follows (using Proposition 2, if necessary) that *X* is in *AC*. Therefore $AB \subset AC$. But by a similar argument (interchanging *B* and *C* in the preceding), $AC \subset AB$. Therefore $AB = AC$.

Now if *D* is any point of *AB* other than *C*, then $AB = AC = CD$. It follows then that if *CD* is any line which contains *A* and *B*, then $AB = CD$. Consequently, *AB* is the only line which contains *A* and *B*.

Corollary 1 Two distinct lines cannot have more than one point in common.

Corollary 2 Given any line *DE*, there is a point *F* not contained in *DE*.

We leave the proofs of the corollaries to the reader.

In the systems of Examples 1 and 2, the segment joining two points may be empty. The following proposition shows that such cannot be the case in a system which satisfies Axioms 1–6.

Proposition 4 *If A and B are distinct points, then there is a point F between A and B.*

Proof The reader might refer to Fig. 5.6 during this proof, although the proof is, of course, not dependent on the figure. By Proposition 3, Corollary

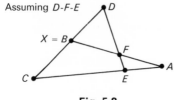

Assuming *D-F-E*

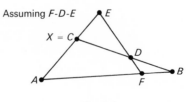

Assuming *F-D-E*

Fig. 5.8   Fig. 5.9

2, there is a point *E* not in *AB*. By Axiom 4 there is *C* such that *A-E-C*; *C* is not in *AB*. For if *C* were in *AB*, we would have *AC* = *AB*, which would make *E* a point of *AB*. Again by Axiom 4, there is *D* such that *B-C-D*. By Axiom 5, then, there is *F* between *A* and *B*.

Corollary *Any line contains at least five points.*

The proof is left as an exercise. Note that this corollary and Proposition 3 tell us that we have a linear space.

In Fig. 5.6 it appears that *E* lies between *D* and *F*. The next proposition states that such is indeed the case.

Proposition 5 *If A, B, and C are three noncollinear points, and D and E are points such that C is in (BD) and E is in (CA), that is, in the situation of Axiom 5, then E is in (DF).*

Proof We outline the proof, leaving the details for the reader. *D*, *E*, and *F* are all in *DE*; hence *D-E-F*, *D-F-E*, or *F-D-E*. Now *E*, *C*, and *D* are noncollinear; this fact and the assumption *D-F-E* leads to both *D-B-C* and *B-C-D* which is impossible. By a similar argument, *F-D-E* is impossible. Figs. 5.8 and 5.9 give further hints as to how the proof should go.

Proposition 6 *If A, B, and C are noncollinear and B-A'-C, C-B'-A, and A-C'-B, then A', B', and C' are also noncollinear* (Fig. 5.10).

Proof If *A'*, *B'*, and *C'* were collinear, then we would have *A'-B'-C'*, *B'-A'-C'*, or *A'-C'-B'*. Suppose *A'-B'-C'*. Now *A'*, *C'*, and *B* cannot be collinear; for, if they were, then the line containing these points would also

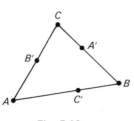

Fig. 5.10

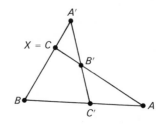

Fig. 5.11

contain A and C (Proposition 3), which contradicts the noncollinearity of A, B, and C. Then by Axiom 5, since B-C'-A and C'-B'-A', there is a point X of AB' between B and A' (Fig. 5.11). Since C is a point of both $A'B$ and AB', it follows that $X = C$. But then we have both B-C-A' and B-A'-C, contradicting Proposition 2. Therefore A'-B'-C' is impossible. The proofs that B'-A'-C' and A'-C'-B' are also impossible are similar and left as exercises.

EXERCISES

1. Prove that the systems defined in Examples 1 and 2 satisfy Axioms 1 through 4.

2. Prove that the system defined in Example 3 satisfies Axioms 1 through 6.

3. Prove the corollaries to Proposition 2.

4. Prove the corollaries to Proposition 3.

5. Prove Proposition 5.

6. In the proof of Proposition 6, prove that B'-A'-C' is impossible.

7. Find an example of a system which satisfies Axioms 1 and 2 in which two lines intersect in more than one point. What additional axioms from Axioms 3–6 must be satisfied before it is certain that two lines can intersect in no more than one point.

8. Try to provide a system which satisfies all of Axioms 1–6 except Axiom k for $k = 1, 2, 3, \ldots, 6$. If you can provide such a system for each k, then you will have shown that Axioms 1–6 are *independent*, that is, no one of these axioms can be proved from the others. Explain why this is the case. If you cannot prove the independence of some Axiom k, try to prove Axiom k from the other axioms.

9. Prove the corollary to Proposition 4.

10. Is the following proof that (AB) contains infinitely many points valid or invalid based on what has been defined and proved in this section? If this proof uses more than we are entitled to use at this point, try to pinpoint precisely what additional statements would have to be proved in order to make the proof valid.

Proof By Proposition 4 there is a point X_1 in (AB). Now each point of (AX_1) is also a point of (AB). By Theorem 4 there is a point X_2 of (AX_1), and X_2 is a point of (AB). Then there is a point X_3 of (AX_2) which is also a point of (AB) as shown in Fig. 5.12. Continuing in like fashion, we can produce a distinct point of (AB) for each positive integer n. Therefore (AB) contains infinitely many points.

Fig. 5.12

5.4 SOME PROPERTIES OF LINES

In this section we will develop some basic properties of lines, in particular the order properties. No axioms beyond Axioms 1 through 6 will be needed for this section.

Proposition 7 If B is in (AC) and C is in (BD), then B is in (AD).

Proof There exists a point O not in AB (Fig. 5.13). There is P in (OB). (See Proposition 4.) By Axiom 5 and Proposition 5, there is Q in (OC) with P in (AQ). Similarly, there is R in (OB) such that Q is in (DR). The points A, Q, and D are noncollinear; for if they were collinear, then P, and hence O, would be in AB, contradicting O not in AB. By Axiom 5, then Q in (RD) and P in (AQ) imply that there exists X in (AD) with X in RP. But RP and AD share B in common. Therefore $X = B$; hence B is in (AD).

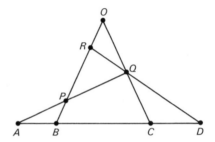

Fig. 5.13

The reader should observe that although Proposition 7 pertains to the internal ordering of AB, its proof requires the use of other lines and points not in AB. This is because of the particular axiomatization we have chosen. Another axiomatization of Euclidean geometry might choose to make more assumptions about the internal structure of lines, rather than accept something like Axiom 5. This is a matter of taste. If one is trying to find a convincing set of axioms for Euclidean plane geometry, the prime requisite is that each axiom express some very basic property of Euclidean geometry.

Proposition 8 If B is in both (AC) and (AD) and $C \neq D$, then either C is in (BD) or D is in (BC).

Proof Since C and D are in AB, $AB = CD$. Since B is in CD, but B is neither C nor D, we have either C is in (BD), or B is in (CD), or D is in (BC). We must therefore show that B cannot be in (CD).

Assume B is in (CD). We may find points O and P not in BC such that C is in (OP), as shown in Fig. 5.14. Since C is in (OP) and B is in (CD), there is Q in (DO) with B in (PQ). Since A is in BC, A is not in CP. Therefore C in (OP) and B in (CA) give a point R in (AO) such that B is in (PR).

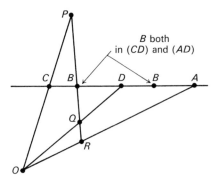

Fig. 5.14

We therefore have points B, Q, and R which must be noncollinear by Proposition 6. But B, Q, and R all lie on BP, a contradiction. Therefore B cannot be in (CD) and the proposition is proved.

The proofs of the following corollaries are left as exercises.

Corollary 1 *If B is a point of both (AC) and (AD) and $C \neq D$, then either C is in (AD) or D is in (AC).*

Corollary 2 *If B is in (AD) and C is in (AD) with $B \neq C$, then either B is in (AC) or C is in (AB).*

Corollary 3 *If B is in (AC) and C is in (AD), then C is in (BD).*

Corollary 4 *If B is in (AC) and C is in (AD), then B is in (AD).*

In an informal view of Euclidean geometry we know that if we remove one point from any line, then we essentially split the line into two sections, or *rays*. These rays are composed of the points which come before the point removed and those which come after the point removed. The following proposition makes this intuitive idea more precise in the geometry we are constructing.

Proposition 9 *If A is any point of the line AB, then the points of AB exclusive of A (that is, the points of $AB - \{A\}$) form two sets such that A is between any point of the first set and any point of the second set, but is not between any two points of the same set.*

Proof Let $/AB = \{X \mid A \text{ is in } (XB)\}$ and

$$AB/ = \{Y \mid Y = B, Y \text{ is in } (AB), \text{ or } B \text{ is in } (AY)\}.$$

Then AB consists of A, $/AB$, and $AB/$.

Suppose that X is in $/AB$ and Y is in $AB/$. Then A is in (XB) (by definition of $/AB$); moreover, we have either $Y = B$ (in which case it follows at once that A is in (XY)), or Y is in (AB), or B is in (AY). If Y is in (AB), then Corollary 3 of Proposition 8 gives A in (XY). If B is in (AY), then Proposition 7 gives A in (XY). We have therefore proved that A is between any point of $/AB$ and $AB/$. We now show that A is not between two points of $/AB$.

Suppose X and X' are points of $/AB$. Then A is in (XB) and also in $(X'B)$. Then by Proposition 8, either X is in (AX') or X' is in (AX). (To apply Proposition 8, interchange A and B in the statement of Proposition 8, and let $C = X$ and $D = X'$.) This, however, precludes the possibility of A being in (XX'). It remains to be shown that A is not between any two points of $AB/$.

Suppose Y and Y' are any two points of $AB/$. Then either Y and Y' are both in (AB), in which case we have either Y in (AY') or Y' in (AY) by Corollary 2 of Proposition 8; or Y is in (AB) and B is in (AY'), in which case we have Y in (AY') by Corollary 4 of Proposition 8; or B is in both (AY) and (AY'), in which case we have either Y in (AY') or Y' in (AY) by Corollary 1 of Proposition 8. In any case, it is impossible to have A in (YY').

Corollary $/AB$ and $AB/$ *share no points in common.*

The proof of the corollary is left as an exercise.

Definition 2 *The two sets of points of Proposition 9 are called* **half-lines** *or* **rays** *(Fig. 5.15). The point A is called the* **end point** *of either ray. Given two distinct points A and B, then the ray of AB which has B as an end point and does not contain A is called the* **prolongation of** (AB) **beyond** B. *The ray which has A as an end point and which contains B will be denoted by $AB/$. The other ray of AB will be denoted by $/AB$.* (Note that a ray does not contain its end point.)

The following are corollaries to Definition 2.

Corollary 1 *If B is a point of a ray which has A as an end point, then the points of this ray exclusive of B are in either (AB) or $/BA$.* (Since the ray itself is $AB/$, this corollary says that

$$AB/ - \{B\} = (AB) \cup /BA.)$$

Prolongation
of (AB)
beyond B **Fig. 5.15**

Proof The set $AB/ - \{B\}$ consists of all points X, with X in (AB) or B in (AX). But if B is in (AX), then X is in $/BA$.

Corollary 2 *If C is a point of (AB), then the points of $(AB) - \{C\}$ are in either (AC) or (BC), but not both. That is, $(AB) - \{C\} = (AC) \cup (BC)$ and $(AC) \cap (BC) = \varnothing$.*

Proof We have B a point of $/CA$, while A is a point of $CA/$ (since C is in (AB)). But since $/CA$ contains all X in (AC), and $CA/$ contains all points of (BC), (AC) and (BC) share no points in common. If X is a point of (AB) other than C, then, by Corollary 2 of Proposition 8, C is in (AX) or X is in (AC). Suppose C is in (AX). Then by Corollary 3 of Proposition 8, C in (AX) and X in (AB) gives X in (BC). Therefore X is either in (BC) or (AC).

We have said before that the lines of Euclidean plane geometry should look like the real number line. The real numbers form a totally ordered set with "less than or equal to" as the total ordering. The geometric properties of the real number line can be expressed in terms of this ordering; for example, if A and B are real numbers, then $(AB) = \{X \mid A < X < B\}$. If the lines of our geometry are to look like the real number line, then we ought to be able to totally order each line and relate the segments, intervals, and rays of the line to the total ordering. We have at this point done all the necessary groundwork to enable us to define a total ordering on each line as follows:

Definition 3 *Let L be any line and A be any point of L. Then A separates L into two rays, r_1' and r_2'. Set*

$$r_1 = r_1' \cup \{A\} \quad \text{and} \quad r_2 = r_2' \cup \{A\} \quad \text{(Fig. 5.16)}.$$

We define $\leq$ by saying:

a) $X \leq Y$ *for every X in r_1 and Y in r_2.*

b) *If X and X' are both in r_1', then $X \leq X'$ if $X = X'$ or if X' is in (XA).*

c) *If Y and Y' are both in r_2', then $Y \leq Y'$ if $Y = Y'$ or if Y is in (AY').*

If $X \leq X'$ but $X \neq X'$, we write $X < X'$. If $X \leq X'$, we may write $X' \geq X$.

The remainder of this section refers to the situation of Definition 3.

Proposition 10 $\leq$ *is a total ordering of L.*

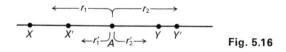

Fig. 5.16

Proof

P1) From (a) we have $A \le A$. From (b) and (c) we have $W \le W$ for any point W of L other than A.

P2) Suppose $W \le W'$ and $W' \le W$ for points W and W' of L. Assume first that W is in r_1 and W' is in r_2. From (a) we also have W' is in r_1 and W is in r_2. But the only point in both r_1 and r_2 is A; hence $W = W' = A$.

Suppose next that W and W' are both in r_2'. If $W \ne W'$, then W is in $(W'A)$ and W' is in (WA). But since W is in $(W'A)$, (WA) is a subset of $(W'A)$ by Corollary 2 of Definition 2; hence W' is a point of $(W'A)$, a contradiction of Axiom 1. Therefore $W = W'$.

We leave the case when W and W' are both in r_1' as an exercise.

P3) Suppose $W \le W'$ and $W' \le W''$. As in proving (P2) we have a number of different cases. The case with W and W' in r_1 and W'' in r_2 is trivial, and the cases where either W or W' is in r_2' and W'' is in r_1 are impossible. Therefore it remains to prove (P3) for the cases when W, W', and W'' are either all in r_1', or all in r_2'. We prove (P3) for the former case and leave the latter case as an exercise.

Assume, then, that W, W', and W'' are all in r_1'. Then W' is in (WA) and W'' is in $(W'A)$. Therefore $(W'A)$ is a subset of (WA). Since W'' is in $(W'A)$, W'' is in (WA); hence $W \le W''$.

T) Suppose W and W' are any points of L. If W is in r_1 and W' is in r_2, then $W \le W'$. Assume W and W' are both in r_1' and $W \ne W'$. Since W, W', and A are collinear, we have A in (WW'), W' in (AW), or W in (AW'). Since W and W' are both in r_1', we cannot have A in (WW'). (See Proposition 9.) Thus, either W' is in (AW), in which case we have $W \le W'$, or W is in (AW'), in which case we have $W' \le W$. Similarly, if W and W' are both in r_2', either $W \le W'$ or $W' \le W$.

This completes the proof that $\le$ is a total ordering of L.

The next proposition characterizes segments, intervals, and rays in relation to the total ordering.

Proposition 11 *Let B and C be points of L with $B < C$. Then:*

a) $(BC) = \{X \mid B < X < C\}$
b) $[BC] = \{X \mid B \le X \le C\}$
c) $BC/ = \{X \mid B < X\}$
d) $/BC = \{X \mid X < B\}$

Proof The proof of (a) is divided into cases according to the location of B and C.

CASE 1 B and C are both in r_1. Suppose X is in (BC). Since $B < C$, C is in (BA). Then by Corollary 3 of Proposition 8 we have C is in (XA); hence $X < C$. Moreover, since (AB) contains (BC), X is in (AB); hence $B < X$. That is, $B < X < C$. Therefore $(BC) \subset \{X \mid B < X < C\}$. It remains to be shown that $\{X \mid B < X < C\} \subset (BC)$. Suppose then that $B < X < C$. Then X is in (AB), C is in (XA), and C is in (AB). Since X, B, and C are collinear, we have B in (XC), C in (XB), or X in (BC). If (XC) contains B, then this together with C in (BA) gives B in (XA) (Proposition 7), contradicting $B < X$. But if C is in (XB), then this, together with C in (AB), gives either X in (AC) or A in (XB) by Proposition 8. X cannot be in (AC) since this contradicts $X < C$, and A cannot be in (XB) since X is in (BC). It follows then that X is in (BC). Consequently, $\{X \mid B < X < C\} \subset (BC)$.

CASE 2 B and C are both in r_2. The proof for this case is left as an exercise.

CASE 3 B is in r_1' and C is in r_2'. By Corollary 2 of Definition 3, $(BC) = (BA) \cup (AC) \cup \{A\}$. Applying Cases 1 and 2 to this equality, we have

$$(BC) = \{X \mid B < X < A\} \cup \{X \mid A < X < C\}.$$

But the right side of this equality is easily seen to be $\{X \mid B < X < C\}$.

The proof of (b) follows immediately from the definition of $[AB]$.

In the proof of (c), by definition, $BC/ = \{X \mid X$ is in (BC), C is in (BX), or $X = C\}$. Now $(BC) \cup \{C\} = \{X \mid B < X \leq C\}$. We now consider those X for which $X = C$ or C is in (BX). If $X = C$, then $B < C$ by hypothesis; therefore let $X \neq C$. Since A, B, and C are collinear, we have A in (BC), C in (BA), or B in (AC). But A in (BC) and C in (BX) would imply A in (BX); hence this would give B in r_1 and X in r_2, which implies $B < X$. Now C in (BA) and C in (BX) gives either X in (CA) or A in (CX); in either case, $B < X$. Also B in (AC) and C in (BX) gives C in (AX) by Proposition 7, from which it follows that $B < X$. We therefore have $BC/ \subset \{X \mid B < X\}$. Suppose now that $B < X$. If $B < X < C$, then from (a) we have X in $(BC) \subset BC/$. If $X = C$, then X is in $BC/$. If $B < C < X$, then again by (a), C is in (BX); hence X is in $BC/$. Therefore $\{X \mid B < X\} \subset BC/$ and (c) is proved.

The proof of (d) is left as an exercise.

Proposition 11 not only tells us that lines have some of the order properties we expect of them, but it also enables us to derive many important facts about lines rather quickly.

Proposition 12 *Suppose A, B, C, and D are distinct and collinear. Then $(AB) \cap (CD)$ is a segment having two of the points A, B, C, and D as end points, or is empty.*

Proof Let L be the line containing A, B, C, and D with a total ordering $\leq$ in accordance with Definition 3. Then one and only one of the following cases occurs.

CASE 1 $A < B < C < D$.   Since $(AB) \cap (CD) = \{X \mid A < X < B\} \cap \{X \mid C < X < D\}$, this intersection is empty. For if X were in this intersection, we would have $C < X$, but $X < B < C$.

CASE 2 $A < C < B < D$. $(AB) \cap (CD) = \{X \mid C < X < B\} = (BC)$.

CASE 3 $D < A < C < B$. $(AB) \cap (CD) = (AC)$.

There are a number of other cases to consider, twenty-one to be exact, but they all work out as easily as the three cases presented.

Proposition 13 *Any segment, and, hence, any ray or line, or prolongation of a segment contains infinitely many points.*

We leave the proof of Proposition 13 as an exercise.

Thus far we have primarily been concerned with the internal structure of a line. In the next chapter we define the term *plane* and begin to investigate its properties.

EXERCISES

1. Prove Corollaries 1 and 2 of Proposition 8.

2. Prove Corollaries 3 and 4 of Proposition 8.

3. Prove the corollary to Proposition 9.

4. In the proof of Proposition 10, prove (P2) for W and W' in r_1'; prove (T) for W and W' in r_1'.

5. Prove (a) of Proposition 11 for Case 2. Prove (d) of Proposition 11.

6. Prove Proposition 13.

7. Refer to Fig. 5.16 when doing the following:
   a) Suppose a point A' of L is chosen from r_1'. Then A' separates L into rays s_1' and s_2' with A in s_1'. Then L can be totally ordered in the manner described in Definition 3 but using A', s_1', and s_2'. Let & be the total ordering thus obtained. Prove that & and $\leq$ are the same total ordering; that is, prove $C \leq D$ if and only if C & D.
   b) Interchange r_1 and r_2 in Definition 3. Let the total ordering obtained by carrying out the procedure of Definition 3, with r_1 and r_2 interchanged, be denoted by %. Prove that $C \leq D$ if and only if D % C.

8. We define a subset U of a line L to be open if given any point A of U, some segment which contains A is a subset of U. Prove each of the following:

a) L and the empty set are both open.
b) The intersection of any two open sets is open.
c) The union of any family of open sets is open.

9. Let S be any set totally ordered by a relation $\leq$. Define s *is between u and v* to mean "$u < s < v$ or $v < s < u$." Which of the axioms do the points of S satisfy?

6 Properties of the Plane. Congruence

6.1 TRIANGLES AND PLANES

Let us pause for a moment to consider what we have in hand so far in our study of Euclidean plane geometry. We have two undefined terms, *point* and *between,* a number of defined terms, among them *line, segment,* and *ray,* six axioms, and various propositions and corollaries. But virtually all of our work thus far has been concerned with lines, either directly, as in Proposition 10, or indirectly, as in Proposition 11. Propositions 10 and 11 can be considered the culmination of our results about lines in Chapter 5, for these propositions tell us that lines can be ordered so that they look just as we would expect. Put another way, we have shown that the pictures we have been drawing of lines all these years have not really been unreasonable. This is not meant to imply that we are through with lines. We have already noted that even the "nice" lines we now have are not necessarily Euclidean lines; for example, they may contain only countably many points, and are not necessarily connected. But we can at least feel confident that we are moving in the right direction.

Now let us look at some of the things we don't have. Distance is something that can be measured in the Euclidean plane, yet we have no way to measure the length of a line segment, or the distance between two points; neither do we have any method of finding the area of a plane figure. As a matter of fact, we do not yet have anything that can properly be called a plane figure. We have not even defined what we mean by *plane*; but since "plane" is not one of our primitive terms, we must either define it, or accept it as yet another primitive term and give its properties by means of additional axioms. We will define it.

What should we use as the starting point for a definition of a plane? The essential fact about any plane is that it is uniquely determined by three noncollinear points. How, given three noncollinear points A, B, and C, can we characterize the plane ABC? We might start with the triangle ABC, which, of course, would be a subset of the plane ABC. We therefore make the following definition.

Definition 1 *Let A, B, and C be three noncollinear points. We define the* **triangle** *ABC to be $\{A, B, C\}$ together with (AB), (AC), and (BC). Triangle ABC may be denoted by $\triangle ABC$. A, B, and C are called the* **vertices** *and (AB), (AC), and (BC) the* **sides** *of $\triangle ABC$.*

If we have a plane (as we usually imagine one) containing $\triangle ABC$, then any point of the plane is collinear with two points of the triangle; moreover, any point not on the plane is not collinear with two points of the triangle. This informal observation inspires the following formal definition.

Definition 2 *Let A, B, and C be three noncollinear points. The* **plane** *ABC is defined to be*

$$\{X \mid X \text{ is collinear with at least two points of } \triangle ABC\}.$$

Plane ABC may be denoted merely by ABC.

It is clear from Definition 2 that $ABC = ACB = BCA = CAB = BAC$. There are, however, fundamental properties of planes which are not clear from Definition 2. For example, we want any three noncollinear points of ABC to uniquely determine ABC; that is, if A', B', and C' are three noncollinear points of ABC, then $A'B'C' = ABC$. We also want any line two of whose points are in a plane to be entirely contained in the plane. We now begin proving some of the basic properties of planes.

The following proposition looks something like Axiom 5. It differs from Axiom 5, however, in that here F is given and the existence of E is proved, while in Axiom 5, E is given and F is assumed to exist.

Proposition 1 *If A, B, and C are three noncollinear points, and D and F are such that C is in (BD) and F is in (AB), then there is E in both (AC) and (DF). (Cf. Fig. 6.1.)*

Proof By Axiom 4 we can find a point O such that B is in (AO). By Corollaries 3 and 4 of Proposition 8 of Chapter 5, it also follows that F is in (AO) and B is in (FO). Since C is in (BD), Proposition 5 (Chapter 5) and Axiom 5 give a point P such that C is in (OP) and P is in (FD). Similarly, from F in (AO) and P in (FD), there is a point Q such that P is in (OQ)

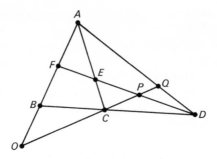

Fig. 6.1

and Q is in (DA). Then P in (OQ) and C in (OP) imply that P is in (CQ) by Corollary 3 of Proposition 8 of Chapter 5. Since A, C, and Q are non-collinear, Q in (AD) and P in (QC) imply by Axiom 5 that there is a point E on DF in (CA). By Axiom 5 and Proposition 2 of Chapter 5, DE intersects (AB) only in F; hence by Proposition 5 of Chapter 5, E is in (DF).

Combining Proposition 1, Definition 1, and Axiom 5, we obtain:

Corollary *A line which meets one side of a triangle and a prolongation of another side meets the third side also.*

The following proposition gives a bit more workable definition of a plane than does Definition 2.

Proposition 2 *If O is any point of side (AB) of $\triangle ABC$, then ABC*

$$\{X \mid X \text{ is in } OW \text{ for some point } W \text{ of } \triangle ABC, W \neq O\}.$$

Proof The complete proof is quite lengthy, and must be broken down into cases. We will begin the proof and prove one case, leaving the proofs of the other cases indicated to the determined reader. According to Definition 2,

$$\{X \mid X \text{ is in } OW \text{ for some point } W \text{ of } \triangle ABC, W \neq O\} \subset ABC.$$

Hence, inclusion must be shown in the other direction. By the corollary to Proposition 1, AB, BC, and AC are all contained in ABC. Suppose X is a point of ABC not contained in AB, BC, or AC. Since X is in ABC, it is collinear with points M and N of $\triangle ABC$; but M and N are not both in (AB), (BC), or (AC).

CASE 1 Suppose one of the points M and N, say M, is A (Fig. 6.2). Since M and N are not in the same side of $\triangle ABC$, N must be in (CB). If X is on either prolongation of (AN), then OX intersects (AB) in some point by

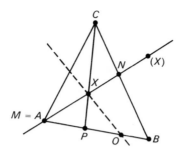

Fig. 6.2

the corollary to Proposition 1. If X is in (AN), then by that same corollary applied to $\triangle ANB$, there is P, such that P is in (AB) and X is in (CP). Now either $O = P$, or O is in (AP), or O is in (PB). If $P = O$, then OX meets $\triangle ABC$ in C. If O is in (AP), then OX meets (CB); while if O is in (PB), then OX meets (AC). If $M = B$, then the same argument works. We have therefore shown that if X is collinear with A or B and a point of $\triangle ABC$, then it is also collinear with O and a point of $\triangle ABC$.

CASE 2 M is in (AB) and N is in (CB).

CASE 3 Neither M nor N is on (AB).

Proposition 3 *If A', B', and C' are any three noncollinear points of ABC, then $A'B'C' = ABC$.*

Proof We first prove that if A' is any point of AB other than B, then $ABC = A'BC$. If $A = A'$, then the statement is trivial. Assume $A \neq A'$; then A is in (AB), A' is in (AB), or B is in (AA').

CASE 1 A is in $(A'B)$. Choose any point O in (AB), as shown in Fig. 6.3; then O is also in $(A'B)$. By Proposition 2, ABC consists of all points collinear with O and some point of $[BC]$ or $[CA]$. But by the corollary to Proposition

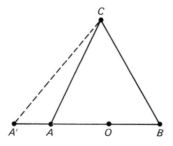

Fig. 6.3

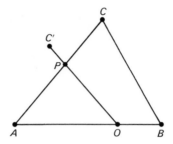

Fig. 6.4

1, this is the same as the set of points which are collinear with O and a point of $[BC]$ or $[A'C]$. Hence in this case $ABC = A'BC$.

The two cases A' in (AB) and B in (AA') are left as exercises. We have, therefore, that if A' is any point of AB other than B, then $ABC = A'BC$. From this it follows that if C' is any point of ABC not on AB, then $ABC = ABC'$. For, let O be any point of (AB). From Proposition 2 we have that OC' meets either (CA) or (CB). Assume OC' meets (CA) in P (Fig. 6.4). By what has already been proved,

$$ABC = ABP = APO = C'AO = C'AB = ABC'.$$

Suppose now that A', B', and C' are three noncollinear points of ABC. Then at least one of these points, say C', is not in AB. Therefore $ABC = ABC'$. A' and B' cannot both be in AC'; hence, suppose B' is not in AC'. Then $ABC' = AB'C'$. But A' is not on $B'C'$; hence $AB'C' = A'B'C'$. Consequently, $ABC = A'B'C'$.

The proofs of the remaining propositions in this section are left as exercises.

Proposition 4 *A line having two points in common with a plane lies entirely in the plane.*

Hilbert takes Proposition 4 as an axiom (Axiom I.6), as do certain other mathematicians who have axiomatized Euclidean geometry. This proposition can also be taken as the definition of a plane, for example, compare Euclid's Definition I, 7.

Corollary *If two planes have two points in common, they have a line in common.*

Proposition 5 *If L is a line in the plane of $\triangle ABC$ and L meets one side of $\triangle ABC$ in exactly one point, then L contains exactly one other point of $\triangle ABC$.* (Remember that according to Definition 1, no side of a triangle contains any vertices of the triangle.)

EXERCISES

1. Prove Proposition 2 for either Case 2 or Case 3.

2. In the proof of Proposition 3, prove that $ABC = A'BC$ for the case A' in (AB), or the case B in (AA').

3. Prove Proposition 4; prove the corollary to Proposition 4.

4. Prove Proposition 5.

5. Let A, B, and C be three noncollinear points. The following terms will be defined in the next section. Before proceeding to the next section, try to compose a suitable definition for each term, based on your previous experience with Euclidean geometry and on what has been done thus far in Part II of this text.
   a) interior of $\triangle ABC$ b) angle ABC
   c) convex subset of a plane   d) polygonal path, or broken line

6. Let A, B, C, and D be four points in the plane ABC, no three of which are collinear. Define the *quadrilateral ABCD*. Discuss some of the basic properties of this quadrilateral in the light of what we know about triangles.

7. Review the definition of a linear variety from Part I. The plane ABC is a linear variety of dimension 2. Prove that any linear variety of dimension 2 in the geometric structure we are considering is a plane.

6.2 SOME TOPOLOGY OF THE PLANE

We might best indicate what we are going to do in this section by stating two of the questions we will consider:

1. In the Euclidean plane as we intuitively envision it, a line divides the plane into two *half-planes*. Does a line in a plane, as we have defined plane in this chapter, also *split* its plane into half-planes? What is the nature of a half-plane?

2. In the Euclidean plane, as intuitively envisioned, a triangle has both an interior and an exterior. Is this true in the plane we have defined? What properties do the interior and exterior of a triangle have?

When we ask a question like, "What is the nature of the set . . . ?", the answer must be in terms that are precisely defined in the system with which we are working. We will define terms to help us describe the sets we are interested in; these terms will be primarily *topological* in nature. The underlying theme of this section is the question: Into how many and what kind of sets do various combinations of lines, rays, etc., divide the plane? We begin with a series of definitions. Throughout this section we assume that everything under consideration is in the same plane.

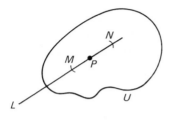

Fig. 6.5

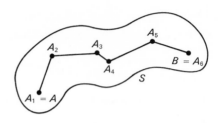

Fig. 6.6

Definition 3 *A subset U of a plane is said to be **open** if, given any point P of U and any line L which contains P, there are points M and N such that P is in (MN) and (MN) is contained in U (Fig. 6.5).*

A set of n − 1 intervals $[A_1A_2]$, $[A_2A_3]$, . . . , $[A_{n-1}A_n]$ *is called a **polygonal path joining** A_1 **and** A_n.* *(A single interval is a special case of a polygonal path.)* *A set S is said to be **polygonally connected** if, given any two points A and B of S, there is a polygonal path joining A and B which lies entirely in S (Fig. 6.6).* *A set S is said to be **convex** if the interval joining any two points of S is a subset of S.*

*A subset V is called a **region** if (1) V is open, and (2) V is polygonally connected.*

The plane and the empty set are both convex and open. Any line, segment, or interval is convex, but is not open. The plane and empty set are also examples of regions.

The next two propositions give important properties of convex sets and open sets.

Proposition 6 *The intersection of any family of convex sets is convex.*

Proof Let $\mathscr{K}$ be a family of convex sets and suppose A and B are points of $\bigcap \{K \mid K$ is in $\mathscr{K}\}$. Then A and B are in K for each K in $\mathscr{K}$. Therefore $[AB]$ is in K for each K in $\mathscr{K}$. Consequently, $[AB]$ is in $\bigcap \{K \mid K$ in $\mathscr{K}\}$; therefore this intersection is convex.

The proof of the next proposition, which uses Proposition 12 of the previous chapter, is left as an easy exercise.

Proposition 7

a) *The plane and the empty set are open.*
b) *The union of any family of open sets is open.*
c) *The intersection of any two open sets is open.*

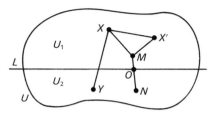

Fig. 6.7

Proposition 7 actually tells us that the open sets form a *topology* for the plane. Although we have not approached geometry in such a way as to make it evident, and have not discussed the notion of a topology, the study of topology, now an important branch of mathematics in its own right, can be considered to be a special branch of geometry.

We now prove our first "separation property" of the plane.

Proposition 8 *Let U be a convex region, and suppose L is any line which contains a point of U. Then the points of $U - L$ form two convex regions U_1 and U_2, such that any segment joining a point of U_1 to a point of U_2 contains a point of L.*

Proof Let O be any point of $U \cap L$. Since U is open, we can find points M and N of U, such that O is in (MN), but (MN) is not a subset of L (Fig. 6.7). Define U_1 to be the set of points of U which can be joined to M by a segment which contains no points of L. We now prove that U_1 is convex and open.

U_1 is convex: Suppose X and X' are two points of U_1. Then if M, X, and X' are noncollinear, the line L cannot meet (XX'). For if it did, then it would also meet either (MX) or (MX'), an impossibility by the way U_1 was defined. Moreover, if W is any point of (XX'), then W is in U_1; for if L met (MW), it would also have to meet (MX), which is impossible. On the other hand, if M, X, and X' are collinear, we may assume X is in (MX'). If L meets (XX'), then L also meets (MX'), contradicting X' as a point of U_1. If W is any point of (XX'), then W is also a point of (MX'); hence, W is in U_1. We have, therefore, shown that if X and X' are any two points of U_1, then (XX') consists entirely of points of U_1; hence, U_1 is convex.

U_1 is open: Let X be any point of U_1 and let L' be any line which contains X. Since X is in U and U is open, there are points M' and N' of L' such that X is in $(M'N')$ and $(M'N')$ is contained in U. A point of $(M'N') - \{X\}$ is either in $(M'X)$ or $(N'X)$. Now L cannot meet both $(M'X)$ and $(N'X)$, for this would imply that $M'N' = L' = L$, an impossibility. If L meets neither $(M'X)$ nor $(N'X)$ and M' and N' are not on L, then M' and N' are both in U_1; since, in such an instance, L would meet none of $(M'X)$,

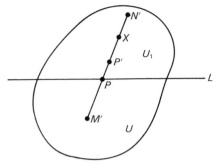

Fig. 6.8

$(N'X)$, and $(M'X)$. Assume, then, that L meets $[M'X]$ in some point P (Fig. 6.8). Choose any point P' in (PX). Then it is easily shown that P' is in U_1; hence $(P'N')$ is contained in U_1 and contains X. Therefore U_1 is open. Consequently, U_1 is a convex region.

Define U_2 to be the set of points Y or U such that (MY) contains a point of L. The proof that U_2 is a convex region is left as an exercise.

From the definitions of U_1 and U_2, it follows that every point of U is either in L, U_1, or U_2, with no overlapping. It remains to be shown that if X is in U_1 and Y is in U_2, then L meets (XY), as shown in Fig. 6.7. Suppose M, X, and Y are noncollinear, and consider $\triangle MXY$. Then L meets (MY); hence L must also meet (XY)—since it cannot meet (MX) because X is in U_1. The case when M, X, and Y are collinear is left as an exercise.

Since the plane itself is a convex region, we have:

Corollary *Any point of the plane not in a line L is contained in one of two convex regions such that a segment joining a point of one of these regions to a point in the other meets L.*

The results of Proposition 8 inspire the following definitions.

Definition 4 *If L is a line, then the regions formed from the points of the plane not in L are called **sides of L**. Each such region is also called a **half-plane**.*

*A set S is said to **separate** two sets T and T' if each polygonal path joining a point of T with a point of T' meets S. A set S is said to **decompose** a region U into regions U_1, U_2, . . . , U_n if each point of U not in S is in one of the regions U_1, . . . , U_n and each pair of the regions U_1, . . . , U_n is separated by $S \bigcup \{X \mid X$ is not in $U\}$.*

As we will see from the next proposition, any line decomposes the plane into two half-planes. We have already shown that the points of the plane

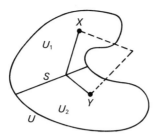

Fig. 6.9

not on the line do not form a convex set, but we have not yet shown that this set is not polygonally connected.

In Fig. 6.9 the set S decomposes the region U into regions U_1 and U_2. This is because

$$S \bigcup \{X \mid X \text{ is not in } U\}$$

separates U_1 and U_2. But S itself fails to separate U_1 and U_2 since we can get from U_1 to U_2 by means of a polygonal path which goes around S.

Any pair of intersecting lines separates the plane into four convex regions. Each of these regions is the intersection of one side of one line with one side of the other. We know each of these intersections is convex and open from Propositions 6 and 7.

Proposition 9 *Any line which contains a point of a convex region decomposes that region into two convex regions.*

We leave the proof of Proposition 9 as an exercise.

EXERCISES

1. Prove Proposition 7.

2. In the proof of Proposition 8 prove that U_2 is a convex region. In the last paragraph of the proof, do the case when M, X, and Y are collinear.

3. Prove Proposition 9.

4. Prove that any pair of intersecting lines separates the plane into four convex regions.

5. A set is said to be *closed* if it is the complement of an open set. Prove that the plane and the empty set are closed. Prove that the intersection of any family of closed sets is closed and the union of any two closed sets is closed. Which of the following sets are closed?

   a) an interval   b) a segment
   c) a line   d) a region

6. If S is any subset, we define the *frontier* of S to consist of all points X such that every open set which contains X contains a point of S and some point not in S. We denote the frontier of S by FrS.
   a) Let $T = \{X \mid X \text{ is not in } S\}$. Prove $FrT = FrS$.
   b) Prove that $S - FrS$ is an open set.
   c) Prove that a set is open if and only if it contains no points of its frontier.

6.3 ANGLES. MORE TOPOLOGY OF THE PLANE

Remember that so far in Part II we have not used any axioms other than Axioms 1 through 6; we shall not need any more axioms for this section. We continue to assume that everything takes place in some plane.

Definition 5 *A point and two distinct rays having that point as a common end point is called an **angle**. The common end point is said to be the **vertex** and the rays the **sides** of the angle. If the rays are collinear, then the angle (which is a line) is called a **straight angle**.*

If a and b are the sides of an angle, we may denote the angle by $\angle ab$. We will use $\angle ABC$ to denote the angle with vertex B and sides $BA/$ and $BC/$.

Proposition 10 *An angle $\angle ABC$ which is not a straight angle decomposes the plane into two regions exactly, one of which is convex. (We call the convex region the **interior** and the other region the **exterior** of the angle.) Moreover, any ray with B as an end point, which contains a point of the interior of $\angle ABC$, meets (AC) and lies entirely in the interior of $\angle ABC$.*

Proof Let a be the side of BA which contains C and a' the side of BC which contains A. Since a and a' are convex regions, $a \cap a'$ is also a convex region (the *interior* of $\angle ABC$). We now show that (1) any ray having B as an end point and containing some point of $a \cap a'$ lies entirely in $a \cap a'$, and (2)

$$\{X \mid X \text{ is not a point of } \angle ABC \text{ or } a \cap a'\}$$

is a region, but is not convex (Fig. 6.10).

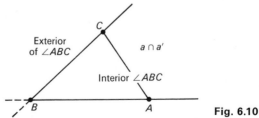

Fig. 6.10

Since $\{X \mid X$ is not a point of $\angle ABC$ or $a \cap a'\}$ is the union of two regions (half-planes), it is a region. We leave it to the reader to prove it is not convex; this establishes (2). We now prove (1).

Suppose O is in $a \cap a'$. If $BO/$ is not a subset of $a \cap a'$, then $BO/$ contains a point Q in the side of BC opposite a', or on the side of BA opposite a. In any case, (BQ) will contain a point of either BC or BA other than B, contradicting the fact that BQ can intersect BC or BA only in B.

It remains to be shown that if O is in $a \cap a'$, then $BO/$ meets (AC). If O is on (AC), then the statement is trivial. Assume O is not in (AC); we distinguish two cases.

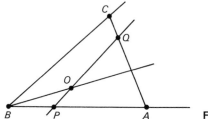

Fig. 6.11

CASE 1 Assume O is on the same side of AC as B. Choose Q in (AC), in Fig. 6.11. Then by Proposition 2, there is a point P of either (AB) or (BC) such that P, Q, and O are collinear. By the corollary to Proposition 1, then, $BO/$ intersects (QA) or (QC), and hence (AC).

CASE 2 Assume O is on the opposite side of AC from B. We leave this case as an exercise.

We now know that angles separate the plane as we would expect them to; we now prove that triangles do the same.

Proposition 11 *A triangle ABC decomposes the plane into two regions exactly one of which is convex. (The convex region is called the* **interior** *and the other region the* **exterior** *of the triangle.) Any ray which has an interior point as its end point meets the triangle in a single point and the interior consists of all points having this property.*

Proof Let x be the side of YZ which contains X and x' be the other side of YZ, where X, Y and Z are A, B, or C and all distinct; thus, a is the side of BC which contains A, while b' is the side of AC which does not contain B. Then $\triangle ABC$ decomposes the plane into the regions $a \cap b \cap c$ and $a' \cup b' \cup c'$, only the first of which is convex. The proofs of these facts are

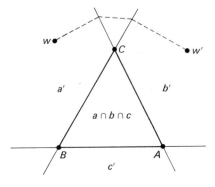

Fig. 6.12

left as exercises, but Fig. 6.12 should prove useful. It is $a \cap b \cap c$, then, that we call the *interior* of $\triangle ABC$. It remains to be shown that the interior of $\triangle ABC$ consists precisely of those points O having the property that any ray with end point O contains exactly one point of $\triangle ABC$.

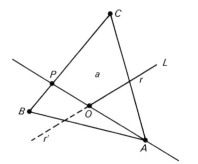

Fig. 6.13

Suppose, then, that O is in $a \cap b \cap c$ and r is a ray with O as its end point (Fig. 6.13). Then r, O, and another ray r' form a line L. Since O is interior to $\angle BAC$, $AO|$ meets (BC) in some point P. Applying the corollary to Proposition 1 to triangles ABP and APC, we see that r and r' each meet $\triangle ABC$ once and only once. Suppose, now, that O is a point such that any ray having O as an end point meets $\triangle ABC$ exactly once (Fig. 6.13). Consider $OA|$ and $|OA$. Then $OA|$ meets $\triangle ABC$ only in A and $|OA$ meets (BC) in a single point P. We can conclude, then, that since O is in (AP), (OA) does not meet (BC); hence O is on the same side of BC as A; that is, O and A are in a. Similarly, O is in b and c; hence O is in $a \cap b \cap c$, the interior of $\triangle ABC$. This completes the proof.

Proposition 12 *If E is a point in the exterior of $\triangle ABC$, then there is a line which contains E but does not meet $\triangle ABC$.*

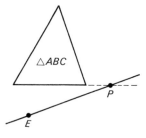

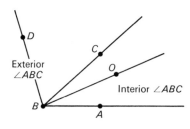

Fig. 6.14   Fig. 6.15

The proof of Proposition 12 is essentially contained in Fig. 6.14. We leave the details to the reader.

The next proposition follows easily from what has come before. Its proof is left as an exercise.

Proposition 13 *If O is an interior point of* $\angle ABC$, *then* $BO/$ *decomposes* $\angle ABC$ *into two convex regions; these regions are the interiors of the angles* $\angle OBC$ *and* $\angle OBA$. *If D is a point of the exterior of* $\angle ABC$, *then* $BD/$ *decomposes the exterior of* $\angle ABC$ *into two regions, at least one of which is convex. One of these regions in the interior or exterior of* $\angle ABD$ *and the other region is the interior or exterior of* $\angle DBC$ (Fig. 6.15).

We intuitively feel that a set of rays having a common end point should divide the plane in a rather nice way. We will soon formalize this rather imprecise feeling. The following definition gives certain terms useful for a discussion of the problem.

Definition 6 *Let U be a region. A set S is said to be a **boundary** of U if* (1) *S contains no points of U,* (2) *any interval joining a point of U with some point not in U meets S, and* (3) *no proper subset of S shares properties* (1) *and* (2).

*Two rays, a and b, are said to be **separated** by* $\angle hk$ *if a, b, h, and k have a common end point and exactly one of a and b is interior to* $\angle hk$ (Fig. 6.16).

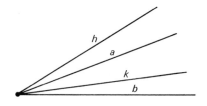

Fig. 6.16

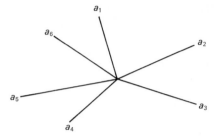

Fig. 6.17

A set of rays $a_1, a_2, \ldots, a_n$ are said to be in the order $a_1 a_2 \ldots a_n$ if no two of the rays are separated by any of the angles $\angle a_1 a_2$, $\angle a_2 a_3, \ldots$, $\angle a_{n-1} a_n$, $\angle a_n a_1$ (Fig. 6.17).

A line is the boundary for each of its sides. A triangle is the boundary for its interior and for its exterior. The sides and vertex of an angle form the boundary for both the interior and exterior of the angle. We leave a formal proof of these statements to the reader.

The following are corollaries of Definition 6.

Corollary 1 *If rays $a_1, \ldots, a_n$ are in the order $a_1 \ldots a_n$, then they are also in the orders $a_2 a_3 \ldots a_n a_1$ and $a_n a_{n-1} \ldots a_2 a_1$.*

Corollary 2 *Any two rays a and b having a common end point are in the orders ab and ba. Any three rays a, b, c having a common end point are in the orders abc, bca, acb, bac, and cba.*

We can now formally say how rays having a common end point decompose the plane.

Proposition 14 *Suppose we are given n rays, $n \geq 2$, having a common end point. Then the rays can be so labeled that they are in the order $a_1 a_2 \ldots a_n$. Moreover, these rays decompose the plane into n regions, of which, at most, one is not convex, and each of which has a set consisting of two of the rays and the common end point as its boundary.*

Proof We proceed by finite induction on n. The proposition is true for $n = 2$. We therefore assume Proposition 14 holds for $n \geq 2$ and prove that this implies its truth for $n + 1$.

Suppose we have a set of $(n + 1)$ rays having a common end point. Choose one ray b from this set. Then the rays exclusive of b form a set of n rays having a common end point. By the induction assumption we can assign notation to these n rays so that they are in the order $b_1 b_2 \ldots b_n$. These k rays then decompose the plane into n regions $U_1, \ldots, U_n$, whose boundaries are $\angle b_1 b_2, \ldots, \angle b_{n-1} b_n, \angle b_n b_1$, respectively. Again by the induction assumption, at most one of these regions is not convex. The other ray b lies in one of these regions, say U_1.

By Proposition 13, b decomposes U_1 into two regions U_1' and U_1'' of which at least one is convex (with both convex if U_1 is convex). Hence, the $n + 1$ rays together decompose the plane into $n + 1$ regions U_1', U_1'', $U_2, \ldots, U_n$ of which at most one is not convex.

The boundary of U_1 is $\angle b_1 b_2$; it follows, then, that the boundaries of the two regions into which U_1 is decomposed by b are $\angle b_1 b$ and $\angle b b_2$. Hence the $n + 1$ rays are in the order $b_1 b b_2 b_3 \ldots b_n$. We may therefore relabel these rays so that they are in the order $a_1 a_2 \ldots a_n a_{n+1}$.

We have come a long way just using Axioms 1 through 6. Yet there are still many concepts related to Euclidean geometry which we cannot yet apply to the plane we have constructed; for example, *congruence*, *distance*, and *perpendicular*, to mention just three. In the next section, we will introduce the concept of *congruence of pairs of points* and begin our discussion of congruence.

EXERCISES

1. Supply the details left to the reader in the proof of Proposition 10.

2. Supply the details left to the reader in the proof of Proposition 11.

3. Prove Proposition 12.

4. Prove Proposition 13.

5. Prove that a line is the boundary for each of its sides. Prove that a triangle is the boundary for its exterior and interior.

6. If S is any subset of the plane, the *convex hull* of S is defined to be the intersection of all convex sets which contain S. Denote the convex hull of S by $C(S)$. Prove that $C(S)$ is nonempty for any nonempty set S. Prove that $C(S)$ is the smallest convex set which contains S. If A, B, and C are three noncollinear points, prove that $C(\{A, B, C\})$ consists of $\triangle ABC$ together with the interior of $\triangle ABC$.

7.* Prove that a set of n distinct lines which meet in some point O decomposes the plane into $2n$ convex regions.

8. Prove that three lines, AB, BC, and CA, which do not all meet in a point decompose the plane into seven regions, one of which is the interior of $\triangle ABC$.

9. Prove that a set of n lines, each pair of which intersect, but no three of which contain a common point, decompose the plane into $(n/2)(n + 1) + 1$ convex regions.

---

* Exercises 7, 8, and 9 are proposed by Veblen.

10. Prove that a line is the frontier (Section 6.2, Exercise 6) of each of its sides. Prove or disprove: The frontier of a region is the same as its boundary.

6.4 CONGRUENCE OF PAIRS OF POINTS

We now introduce a new primitive relation *is congruent to* between ordered pairs of points (of some plane). Intuitively, (A, B) is congruent to (C, D) if the distance from A to B is the same as the distance from C to D. However, we do not yet have a means of measuring distance, and when we introduce a distance into the plane, it will be done by using congruence. Since *is congruent to* is undefined, any interpretation is valid which is consistent with Axioms 1 to 6 and the axioms below. We will denote *is congruent to* by the symbol $\cong$.

Axiom 7 *If* $A \neq B$, *then on any ray with* C *as an end point there is exactly one point* D *such that* $(A, B) \cong (C, D)$.

Axiom 7 corresponds to Euclid's Third Postulate: It shall be possible to draw a circle with a given center through a given point.

Axiom 8 *If* $(A, B) \cong (C, D)$ *and* $(C, D) \cong (E, F)$, *then* $(A, B) \cong (E, F)$. *That is*, ***is congruent to*** *is a* ***transitive*** *relation.*

Axiom 9 *If* $(A, B) \cong (A', B')$ *and* $(B, C) \cong (B', C')$ *and* B *is in* (AC) *and* B' *is in* $(A'C')$, *then* $(A, C) \cong (A', C')$. (See Fig. 6.18.)

Axiom 8 corresponds to the Euclidean axiom: Things equal to the same thing are equal to each other. Axiom 9 corresponds to: If equals are added to equals, then the results are equal.

Axiom 10 $(A, B) \cong (B, A)$.

The structure defined in Example 3 of Chapter 5 satisfies Axioms 1 to 6. We now show that this same structure fails to satisfy Axiom 7.

Example 1 The situation is that of Example 3 of Chapter 5, with (A, B) congruent to (C, D) if the distance from A to B is the same as the distance from C to D (distance measured in the usual analytic geometric way). There is no point D of Y on the line with equation $y = x$ such that $((0, 0), D) \cong ((0, 0), (0, 1))$. For D would have to have coordinates $(1/\sqrt{2}, 1/\sqrt{2})$ or $(-1/\sqrt{2}, -1/\sqrt{2})$ and neither of these are points of Y.

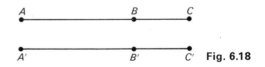

Fig. 6.18

Axioms 7 to 10 carry certain implications about the extent of lines and the unboundedness of the plane. We try to bring this out in the following example.

Example 2 Let W be the set of points in the coordinate plane R^2 inside the circle of radius 10 with center $(0, 0)$. Let *is between* and *is congruent to* be defined as in the example above. Then the structure on W satisfies all of the axioms except Axiom 7. For example, there is no point P on the ray $(5, 0)(6, 0)/$ such that $(P, (5, 0)) \cong ((0, 0), (9, 0))$. The line $(5, 0)(6, 0)$ simply does not extend far enough to make it possible for such a point to exist.

This example also helps point up the relationship between Axiom 7 and the corresponding axiom of Euclid. The circle with center $(5, 0)$ and radius 9 cannot be completed in W because W is not large enough.

We now proceed to investigate the properties of congruence.

Proposition 15 *If B is in (AC) and C' is a point of $A'B'/$ such that $(A, B) \cong (A', B')$ and $(A, C) \cong (A', C')$, then B' is in $(A'C')$.*

Proof By Axiom 7 there is C'' of $/B'A'$ such that $(B, C) \cong (B'. C'')$. (See Fig. 6.19.) Then B' is in $(A'C'')$. (Why?) By Axiom 9 $(A, C) \cong (A', C'')$. Therefore, since C' and C'' are both points of $A'B'/$ such that (A, C) is congruent to both (A', C') and (A', C''), it follows from Axiom 7 that $C' = C''$. Therefore B' is in $(A'C')$.

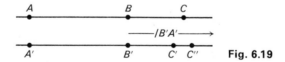

Fig. 6.19

Proposition 16 *If $(A, B) \cong (A', B')$ and $(A, C) \cong (A', C')$, and if C is in $AB/$ and C' is in $A'B'/$, and if $B \neq C$, then $(B, C) \cong (B', C')$.*

Proof If C is in $AB/$ and if $B \neq C$, then either C is in (AB), or B is in (AC). Assume that B is in (AC). Then by the previous proposition, we have B' in $(A'C')$. By Axiom 7 there is C'' on $B'C'/$ such that $(B, C) \cong (B', C'')$. By Axiom 9 applied to A, B, C, and A', B', C'', $(A, C) \cong (A', C'')$. But C' is a point of $A'B'/$ such that $(A, C) \cong (A', C')$; moreover, it is the only such point. Therefore $C' = C''$ and $(B, C) \cong (B', C')$.

Assume now that C is in (AB). Then interchanging B and C in Proposition 15, we find C' is in $(A'B')$. By an argument similar to that of the paragraph above, we find $(B, C) \cong (B', C')$.

The proofs of the following two propositions are left as exercises.

Proposition 17 $(A, B) \cong (A, B)$. (*That is, $\cong$ is a **reflexive** relation.*)

The corresponding statement in Euclid is: Things that coincide are equal. Propositions 17 and 18, together with Axiom 8, tell us that *is congruent to* defines an *equivalence relation* on the set of ordered pairs of points of the plane.

Proposition 18 If $(A, B) \cong (C, D)$, then $(C, D) \cong (A, B)$. That is, *is congruent to* is *symmetric*.

Corollary If $(A, B) \cong (C, D)$ and $(A, B) \cong (E, F)$, then $(C, D) \cong (E, F)$.

We have already mentioned that Euclidean geometry is concerned with properties preserved by rigid motions. Until now we have not been in a position to make the idea of rigid motion precise. Intuitively, a rigid motion between two figures F and F' is a correspondence between the points of F and the points of F' such that the distance between two points of F is the same as the distance between the two points of F' to which they correspond. More formally, in terms of congruence, we have the following definition.

Definition 7 A **plane figure**, or simply a **figure**, is any subset of the plane. Two figures F and F' are said to be **congruent** if there exists a one-one function f from F onto F' such that for any two points A and B of F, $(A, B) \cong (f(A), f(B))$. Such a function f is said to be a **congruency**, or **rigid motion** from F to F'. If F is congruent to F', we write $F \cong F'$.

Example 3 Let R^2 be the usual coordinate plane with its standard analytic geometric structure and congruence of ordered pairs of points as interpreted earlier. Then a rigid motion of R^2 is a one-one function f from R^2 onto R^2 which preserves distance, that is, $D(A, B) = D(f(A), f(B))$ for any two points A and B of R^2 . Any translation and rotation of R^2 is a rigid motion of R^2 . An arbitrary dilatation is generally not a rigid motion. The reader might try to answer the following questions: Is any distance-preserving function from R^2 into itself a rigid motion; that is, is distance preserving sufficient to ensure one-one and onto? Is distance preserving sufficient to ensure that the function is a collineation of R^2 ; in other words, are there rigid motions which are not collineations?

We would expect *is congruent to* to define an equivalence relation on the set of plane figures just as it did on the set of ordered pairs of points. The next proposition fulfills this expectation.

Proposition 19

a) *Any figure is congruent to itself.*

b) *If F and F' are figures such that $F \cong F'$, then $F' \cong F$.*

c) *If F, F', and F'' are figures such that $F \cong F'$ and $F' \cong F''$, then $F \cong F''$.*

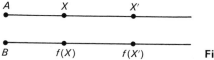

Fig. 6.20

Proof Proposition 19 follows easily once the following facts have been established: The identity function is a congruency. If f is a congruency from F onto F', then f^{-1} is a congruency from F' onto F. And the composition of two congruencies is a congruency. We prove the second of these assertions and leave the proofs of the other two to the reader.

Since f is a one-one function from F onto F', f^{-1} is a one-one function from F' onto F. Let A and B be any two points of F'. Since f is a congruency

$$\left(f^{-1}(A), f^{-1}(B)\right) \cong \left(f(f^{-1}(A)), f(f^{-1}(B))\right) = (A, B).$$

Therefore $(A, B) \cong \left(f^{-1}(A), f^{-1}(B)\right)$; hence f^{-1} is also a congruency.

Proposition 20 *Any point is congruent to any point, any line to any line, any ray to any ray, and any straight angle to any straight angle.*

Proof We will prove that if r and r' are two rays, then $r \cong r'$, and leave the remainder of the proof to the reader. We define a function f from r onto r' as follows: For any point X of r, let $f(X)$ be the unique point of r' such that $(A, X) \cong (B, f(X))$, as shown in Fig. 6.20. It follows at once from Axiom 7 that f is one-one and onto. Suppose X and X' are any two points of r. Then $(A, X) \cong (B, f(X))$ and $(A, X') \cong (B, f(X'))$. By Proposition 16, then $(X, X') \cong \left(f(X), f(X')\right)$. Therefore f is a congruency from r onto r'; hence $r \cong r'$.

Proposition 21 *If $(A, B) \cong (C, D)$, then $(AB) \cong (CD)$ and $[AB] \cong [CD]$.*

Proof Let X be any point of (AB), and Y be the unique point of $CD/$ such that $(A, X) \cong (C, Y)$. By Proposition 15, Y is in (CD). By Proposition 16, $(B, X) \cong (D, Y)$. Set $f(X) = Y$. Then f is a congruency from (AB) onto (CD). To get a congruency from $[AB]$ onto $[CD]$, extend f by setting $f(A) = C$ and $f(B) = D$.

EXERCISES

1. Complete the proof of Proposition 19.
2. Prove Propositions 17 and 18.
3. Complete the proof of Proposition 20.
4. In the proof of Proposition 15, prove that B' is in $(A'C'')$.

5. Refer to Example 3. Is any distance-preserving function from R^2 into R^2 necessarily one-one and onto? Is such a function necessarily a collineation? [*Hint:* Three points A, B, and C are such that B is between A and C if and only if $D(A, B) + D(B, C) = D(A, C)$.]

6. Prove any of the statements below that are correct. If a statement is incorrect, try to find a correct statement as close to it as possible and prove the correct statement.

   a) If $(A, B) \cong (B, C)$ and A, B, and C are collinear, then B is in (AC).
   b) If $(A, B) \cong (A', B')$ and $(B, C) \cong (B', C')$, then $(A, C) \cong (A', C')$.
   c) If $(A, B) \cong (B, C)$ and $AB = BC$, then B is in (AC).
   d) Given (AC), there is a unique point B of (AC) such that $(A, B) \cong (B, C)$.

7. Given any ray r and a segment (AC), prove that r contains infinitely many segments congruent to (AC), no two of which contain a point in common.

8. Prove that the structures defined in Examples 1 and 2 satisfy all of the axioms for Euclidean geometry introduced so far, except Axiom 7.

6.5 CONGRUENCE OF ANGLES

Axioms 7 to 10 ensure that congruence of ordered pairs of points has sufficient properties to make congruence come out all right as far as lines and subsets of lines are concerned. The following axiom is needed to make congruence work for sets of points which are not all collinear.

Axiom 11 *If* A, B, *and* C *are three noncollinear points, and if* D *is a point such that* C *is in* (BD), *and if* A', B', *and* C' *are three noncollinear points with* D' *a point such that* C' *is in* $(B'D')$, *and if*

$$(A, B) \cong (A', B'), \qquad (B, C) \cong (B', C'), \qquad (C, A) \cong (C', A'),$$

and

$$(B, D) \cong (B', D'),$$

then

$$(A, D) \cong (A', D').$$

(See Fig. 6.21.)

The following example points out what can happen without Axiom 11.

Example 4 Let R^2 be the coordinate plane with the linear structure defined in Example 2 of Chapter 4. We will say B is between A and C if B divides A from C on AC, that is, if B is in (AC) with (AC) the line segment joining A and C in accordance with the modified geometry on R^2. We define $(A, B) \cong (C, D)$ if the distance from A to B is the same as the distance from C to D, where the distance from A to B is found by computing the length of (AB). (Note that whereas (AB) is straight in the usual geometry of R^2, (AB) may be bent in the modified geometry; hence the modified distance from A to B may be greater than the distance from A to B computed in the

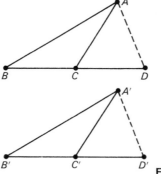

Fig. 6.21

usual fashion.) The reader should verify that the structure thus obtained on R^2 satisfies Axioms 1–10. But it does not satisfy Axiom 11. For let A, B, C, D, A', B', C', and D' be positioned as in Fig. 6.22 with all of the hypotheses of Axiom 11 satisfied. Because of the way distance is measured, (A, D) is generally not congruent to (A', D'). Therefore Axiom 11 is not satisfied.

We are now able to prove some of the well-known congruence results of Euclidean geometry.

Proposition 22 *Two angles* $\angle BAC$ *and* $\angle MON$ *are congruent in such a way that A corresponds to O if there are points P and Q on OM| and ON| such that*

$$(A, B) \cong (O, P), \quad (A, C) \cong (O, Q), \quad and \quad (B, C) \cong (P, Q).$$

(This corresponds to the Euclidean Proposition 8, Book One: If two triangles have two sides of one congruent to two sides of the other, and their third sides are also congruent, then the angle which is contained by the two sides

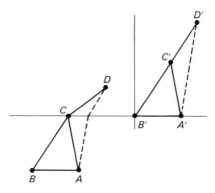

Fig. 6.22

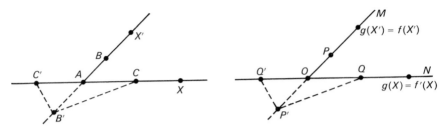

Fig. 6.23

of one triangle is congruent to the angle contained by the two sides of the other.)

Proof By Proposition 20 there are congruencies f from $AB|$ onto $OM|$ and f' from $AC|$ onto $ON|$. We define a function g from $\angle BAC$ into $\angle MON$ as follows:

$$g(X) = \begin{cases} f(X) \text{ for } X \text{ in } AB|, \\ O \text{ if } X = A, \\ f'(X) \text{ for } X \text{ in } AC|. \end{cases}$$

Since f and f' are one-one and onto, g is one-one and onto. Suppose X and X' are two points of $\angle BAC$. If X and X' are either both in $AB| \cup \{A\}$ or both $AC| \cup \{A\}$, then $(X, X') \cong (g(X), g(X'))$, by Proposition 20. Therefore suppose that X is in $AC|$ and X' is in $AB|$ (Fig. 6.23).

Let B' and P' be points of $|AB$ and $|OP$, respectively, such that $(A, B') \cong (O, P')$. Since $(A, C), (C, B), (B, A),$ and (B, B') are congruent respectively to $(O, Q), (Q, P), (P, O),$ and (P, P'), where $(B, B') \cong (P, P')$ by Axiom 9, we have by Axiom 11 that

$$(B', C) \cong (P', Q).$$

Let C' and Q' be points of $|AC$ and $|OQ$, respectively, such that

$$(A, C') \cong (A, C) \qquad \text{and} \qquad (O, Q) \cong (O, Q').$$

Since we have $(A, C), (A, B'), (C, B')$ and (C, C') congruent respectively to $(O, Q), (O, P'), (Q, P'),$ and (Q, Q'), Axiom 11 gives us

$$(C', B') \cong (Q', P').$$

Since we have $(A, C'), (A, B'), (C', B'),$ and (C', X) respectively congruent to $(O, Q'), (O, P'), (Q', P')$ and $(Q', g(X))$, we have by Axiom 11 that

$$(X, B') \cong (g(X), P').$$

Axiom 11, together with (A, X), (X, B'), (A, B'), and (B', X') respectively congruent to $(O, g(X))$, $(g(X), P')$, (P', Q'), and $(P', g(X'))$ gives

$$(X, X') \cong (g(X), g(X')).$$

Therefore g is a congruency and $\angle BAC \cong \angle MON$.

Definition 8 Consider $\angle BAC$. Then the angle with vertex A and sides $AB|$ and $|AC$ is called the **supplement** of $\angle BAC$. If B' is in $|AB$, and C' is in $|AC$, then $\angle BAC$ and $\angle B'AC'$ are said to be **vertical**. (This definition corresponds exactly to the customary notions of supplementary and vertical angles.)

Proposition 23

a) *Supplements of congruent angles are congruent.*
b) *Vertical angles are congruent.*

Proof a) Suppose that $\angle B'A'C' \cong \angle BAC$ by a congruency f such that

$$f(A) = A', \qquad f(B) = B', \qquad \text{and} \qquad f(C) = C'.$$

Let D and D' be points of $|AB$ and $|A'B'$, respectively, such that $(D, A) \cong (D', A')$, as shown in Fig. 6.24. Then by Axiom 11, $(D, C) \cong (D', C')$. Consequently, by Proposition 22, $\angle DAC$, the supplement of $\angle BAC$, is congruent to $\angle D'A'C'$, the supplement of $\angle B'A'C'$.

The proof of (b) is left as an exercise.

Statement (a) of Proposition 23 is Proposition 15 of Euclid's Book One. Euclid would have proved (b) as follows: $\angle DAB \cong \angle D'A'B'$ (any two straight angles are equal), and $\angle BAC \cong \angle B'A'C'$. By the principle of "equals subtracted from equals give equals," we have $\angle DAC \cong \angle D'A'C'$.

After the next definition we will be ready to prove the three most important propositions concerning congruence of triangles.

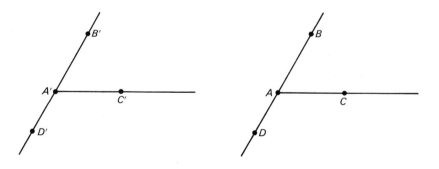

Fig. 6.24

Definition 9 In $\triangle ABC$ the sides (AB) and (BC) are said to **include** $\angle BAC$. The side (AC) is said to be **opposite** $\angle ABC$. Sides (AB) and (BC) are said to be **adjacent** to each other and to $\angle ABC$.

Proposition 24 (s.s.s congruence) *If the sides of one triangle are respectively congruent to the sides of another triangle, then the triangles are congruent* (Euclid, Corollary to Proposition 8 of Book One).

Proof Consider triangles ABC and $A'B'C'$ with (AB), (BC), and (CA) respectively congruent to $(A'B')$, $(B'C')$, and $(C'A')$. There are congruencies f_1, f_2, and f_3 from $AB| \cup \{A\}$ onto $A'B'| \cup \{A'\}$, $BC| \cup \{B\}$ onto $B'C'| \cup \{B'\}$, and $CA| \cup \{C\}$ onto $C'A'| \cup \{C'\}$, respectively. Define the function f on $\triangle ABC$ as follows:

$$f(X) = \begin{cases} f_1(X) \text{ if } X \text{ is in } AB| \cup \{A\}, \\ f_2(X) \text{ if } X \text{ is in } BC| \cup \{B\}, \\ f_3(X) \text{ if } X \text{ is in } CA| \cup \{C\}. \end{cases}$$

To prove f is a congruency from $\triangle ABC$ onto $\triangle A'B'C'$, we must prove (1) f is a function from $\triangle ABC$ onto $\triangle A'B'C'$; (2) f is one-one and onto; and (3) $(X, X') \cong (f(X), f(X'))$ for any X, and X' in $\triangle ABC$. Statements (1) and (2) follow easily from the definition of f. Statement (3) follows from the definition of f if X and X' are both on the same ray and from Proposition 22 if X and X' are on different rays. The details of the proof are left to the reader.

Proposition 25 (s.a.s. congruence) *If two sides and the included angle of one triangle are congruent to two sides and the included angle of another triangle in such a way that the vertex of one angle corresponds to the vertex of the other angle, then the triangles are congruent.*

Proof Consider triangles ABC and $A'B'C'$ with $(AB) \cong (A'B')$, $(BC) \cong (B'C')$, and $\angle ABC \cong \angle A'B'C'$. Suppose f is the congruency between $\angle ABC$ and $\angle A'B'C'$ such that $f(B) = B'$. From Axiom 7 and the fact that $(AB) \cong (A'B')$, it follows that $f(A) = A'$. Similarly, $f(C) = C'$. But then, $(A, C) \cong (f(A), f(C)) \cong (A', C')$. Hence, $(AC) \cong (A'C')$. Therefore, the three sides of $\triangle ABC$ are respectively congruent to the three sides of $\triangle A'B'C'$. By Proposition 24, then, $\triangle ABC \cong \triangle A'B'C'$.

Proposition 25 appears in Euclid as Proposition 4 of Book One before the proposition which corresponds to Proposition 24.

We will wait until the next section to prove a.s.a. congruence. We close this section with another familiar theorem of Euclidean geometry.

Proposition 26 *If two sides of a triangle are congruent, then the angles opposite those sides are congruent.*

Proof Consider $\triangle ABC$ and suppose $(AB) \cong (AC)$. Since (A, B), (B, C), and (A, C) are respectively congruent to (A, C), (B, C), and (A, B), we have from Proposition 22 that $\angle ACB \cong \angle ABC$.

EXERCISES

1. Complete the proof of Proposition 24.

2. Prove (b) of Proposition 23.

3. In Fig. 6.22 of Example 4 let

$$B' = (0, 0), \qquad C' = (\sqrt{3}/2, \tfrac{1}{2}), \qquad D' = (\sqrt{3}, 1),$$
$$A' = (-\sqrt{3}/2, 0), \qquad A = (0, -\tfrac{1}{2}), \qquad B = (-\sqrt{3}/2, -\tfrac{1}{2}),$$
$$C = (0, 0), \qquad \text{and} \qquad D = (\tfrac{1}{2}, \sqrt{3}/2).$$

   Prove that the hypotheses of Axiom 11 are satisfied but not its conclusion.

4. Prove that the structure given in Example 4 satisfies Axioms 1–10.

5. a) Prove that if A, B, and C are distinct noncollinear points and A', B', and C' are collinear with B' in $(A'C')$ such that (A, B) and (B, C) are respectively congruent to (A', B') and (B', C'), then there is no congruency f from $\triangle ABC$ to $[A'C']$ such that

$$f(A) = A', \qquad f(B) = B', \qquad \text{and} \qquad f(C) = C'.$$

   b) Prove that if $\angle ABC$ and $\angle MON$ are not straight angles and $\angle ABC \, \angle MON$, then there are essentially only two congruencies from $\angle ABC$ to $\angle MON$ such that B corresponds to O. Prove an analogous statement if the angles are straight angles.

6. In the proof of Proposition 26, find an explicit congruency between $\angle ACB$ and $\angle ABC$.

7. Discuss as fully as you can the questions raised in each of the following:

   a) Are we justified (on the basis of what we have proved so far) in saying that if $\triangle ABC \cong \triangle A'B'C'$, then any congruency from one triangle to the other takes vertices to vertices?

   b) Define a *right angle* to be an angle which is congruent to its supplement. Can we be certain yet that right angles exist?

7 Circles.
Perpendicular
and Parallel
Lines. Length

7.1 CIRCLES

We continue to assume that everything takes place in some plane unless specifically stated otherwise. We begin this section by introducing the notion *circle*.

Definition 1 *Given distinct points O and A, we define the **circle** with **center** O and **radius** $[OA]$ to be*

$$\{X \mid (O, X) \cong (O, A)\}.$$

*If X is any point of the circle, then $[OX]$ is said to be a **radius**. Two collinear radii of the circle are said to form a **diameter**. The points contained in some radii of the circle but not in the circle itself are said to form the **interior** of the circle. The points not on any radius of the circle are said to form the **exterior** of the circle.*

Circles play an important part in many of the constructions of Euclidean geometry. For example, consider the following proposition and its proof from Euclid's Book One.

Euclid's Proposition 22, Book One *To construct a triangle of which the sides shall be equal (congruent) to three given segments, any two of which together are greater than the third.*

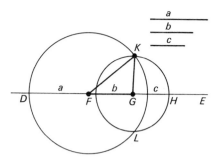

Fig. 7.1

Proof Let *a*, *b*, and *c* be the three given segments, of which any two together are greater than the third . . . : It is required to make a triangle of which the sides shall be congruent to *a*, *b*, and *c*.

Construction (Fig. 7.1) Take a straight line *DE* ending at *D*, but unlimited towards *E*, and make *DF* congruent to *a*, *FG* congruent to *b*, and *GH* congruent to *c*. With center *F* and radius *FD*, describe the circle *DKL*. With center *G* and radius *GH*, describe the circle *HKL*, and let it cut the former circle at *K*. Join *KF*, *KG*. The triangle *KFG* shall be drawn as required.

Proof Because *F* is the center of circle *DKL*, *FD* = *FK*. But *FD* = *a*; therefore *FK* = *a*. Again, because *G* is the center of circle *HLK*, *GH* = *GK*. But *GH* = *c*; hence *GK* = *c*. Also *FG* = *b*. Therefore the three segments *KF*, *FG*, and *GK* are equal to the three, *a*, *b*, *c*. Therefore the triangle *KFG* has its three sides *KF*, *FG*, and *GK* equal to the three given segments, *a*, *b*, *c*.

Although the above excerpt from Euclid points up the greater precision of the techniques we are using over Euclid's, we have primarily introduced this excerpt because it uses circles to obtain a desired result. If, indeed, we want to make use of constructive "straightedge-and-compass" proofs in our own work, we must have some axiom which tells us that circles are in a "proper fashion." If Euclid had not been certain that the circles he used in his proof intersected, his conclusion would not have followed; as a matter of fact, one of the postulates Euclid implicitly assumed, but never stated, is that circles intersect in a "nice" way. We will adopt such an axiom explicitly.

Axiom 12 *A circle which contains a point A interior and a point B exterior to another circle has at least one point in common with the other circle on each side of AB* (Fig. 7.2).

Before using Axiom 12, we make a three-part definition.

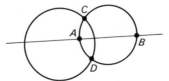

Fig. 7.2

Definition 2 *A triangle is said to be **isosceles** if two of its sides are congruent and **equilateral** if all of its sides are congruent.*

*A **closed half-plane** is a half-plane together with its boundary line.*

Our first constructive proof is, appropriately enough, Euclid's Proposition 1 of Book One.

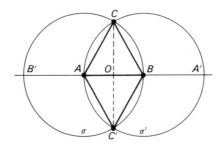

Fig. 7.3

Proposition 1 *If* (AB) *is any segment, then there exists an equilateral triangle of which* (AB) *is a side in any closed half-plane of which AB is a boundary.*

Proof Let σ and σ' be circles with centers A and B, respectively, and each with radius $[AB]$, as shown in Fig. 7.3. Let A' be the point of $/BA$ with $(A, B) \cong (A', B)$. Then $[A'B]$ is a radius of σ' and $[AA']$ is a diameter of σ'. It is easily shown that $[AA']$ contains all of the points of AB which are interior to σ'. Let B' be the point of $/AB$ such that $(B', A) \cong (A, B)$. Now B' is not in $[AA']$, but B' is in AB; therefore B' is exterior to σ'. Circle σ therefore contains a point B interior and a point B' exterior to σ'. Therefore σ and σ' intersect in points C and C', one on each side of AB. Clearly, $\triangle ABC \cong \triangle ABC'$, and these triangles are in different closed half-planes relative to AB.

Proposition 2 *If* (AB) *is any segment, then there is a unique point O of* (AB) *such that* $(A, O) \cong (O, B)$. *We call this point O the **mid-point** of* (AB) *and of* $[AB]$. (The analogous proposition of Euclid is Proposition 10 of Book One.)

Proof We use the situation set up in the proof of the previous proposition (Fig. 7.3). Since C and C' are on opposite sides of AB, (CC') meets AB in a point O. Since (C, B), (C', B), and (C, C') are congruent respectively to (C, A), (C', A), and (C, C'), $\angle BCO \cong \angle ACO$ by some congruency f for which $f(C) = C$. Since $(C, O) \cong (C, O)$, it follows that $f(O) = O$. Since we must also have $f(B) = A$, it follows $(A, O) \cong (B, O)$. We now show that O is in (AB) and is the only point of AB for which $(A, O) \cong (B, O)$.

Suppose that O' is a point of AB other than O for which $(AO') \cong (O'B)$. If O' were in $/BA$, then we would have two points A and B on $O'B/$ such that $(O', B) \cong (O', A)$, contradicting Axiom 7. Therefore O' cannot be in $/BA$; similarly, O' cannot be in $/AB$. Therefore O' must be in (AB). Now the same argument that shows O' must be in (AB) shows that O must be in (AB). Since O and O', then, are both in (AB), we have either O' is in (AO) or O' is in (OB). Suppose O' is in (OB). Then, since $(A, O) \cong (A, O')$ and $(B, O) \cong (B, O')$, it follows from Proposition 15 of Chapter 6 that O is in (BO'), a contradiction to O' in (OB). Therefore O' cannot be in (OB); but, similarly, O' cannot be in (AO). Therefore O is the only point of AB such that $(A, O) \cong (B, O)$, and O is a point of (AB).

Proposition 3 *Consider* $\triangle ABC$. *There is no point* C' *distinct from* C, *on the same side of* AB *as* C, *such that* $(A, C) \cong (A, C')$ *and* $(B, C) \cong (B, C')$.

Proof By Axiom 7, C' cannot lie on either AC or BC. Suppose C' exists and is not in either AC or BC. We distinguish two cases according to whether CC' does or does not meet (AB).

CASE 1 Assume CC' meets (AB) in some point P (Fig. 7.4). Then $(B, C) \cong (B, C')$, $(C, A) \cong (C', A)$, and $(A, B) \cong (A, B)$. Therefore $\triangle ABC \cong \triangle ABC'$ by a congruency f such that $f(C) = C'$ and $f(P) = P$. Consequently, C and C' are both points of $PC/$ such that $(P, C) \cong (P, C')$, contradicting Axiom 7.

CASE 2 Assume CC' does not meet (AB), as shown in Fig. 7.5. Then A and B are on the same side of CC'. Let O be the midpoint of (CC') and P

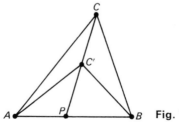

Fig. 7.4

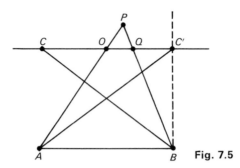
Fig. 7.5

be any point of $/OA$ in the interior of $\angle CBC'$. (The proof that such a point exists is left as an exercise.) Now P is on the opposite side of CC' from B. This, together with Proposition 10 of Chapter 6, enables us to show that (PB) meets (CC') in some point Q.

Since (O, C), (C, A), (A, O), and (A, P) are respectively congruent to (O, C'), (C', A), (A, O), and (A, P), it follows from Axiom 11 that $(C, P) \cong (C', P)$. Therefore, $\angle CBP \cong \angle C'BP$ by a congruency which takes B to B, C to C', and Q to Q. Consequently, $(C, Q) \cong (C', Q)$. But this means that Q is a midpoint of (CC') distinct from O, a contradiction of Proposition 2.

Corollary *If* $\angle ABC \cong \angle ABC'$ *in such a way that* A *corresponds to itself and* C' *is on the same side of* AB *as* C, *then* $BC/ = BC'/$.

The proof is left as an exercise.

We have proved a fair number of results about what happens in any one plane, but we have not yet proved any result which enables us to say that any two planes are geometrically equivalent. That is, although we know that any two planes share many properties in common, we do not know that any two planes are congruent. Until we prove this fact, we cannot be sure that plane geometry does not depend on the particular plane we happen to consider. The following proposition states that all planes are congruent; hence, plane geometry can be studied in any one plane and the results will be valid in all planes.

Proposition 4 *If* ABC *and* $A'B'C'$ *are any two planes, then they are congruent in such a way that* B *corresponds to* B', $BC/$ *corresponds to* $B'C'/$, *and the half-plane containing* A *and bounded by* BC *corresponds to the half-plane containing* A' *and bounded by* $B'C'$.

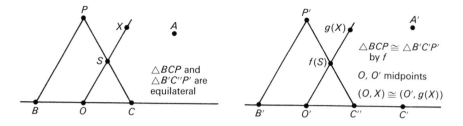

Fig. 7.6

Proof The complete proof is quite lengthy, although straightforward. Fig. 7.6 indicates how a congruency g from ABC onto $A'B'C'$ is constructed. We leave the details to the reader.

The proof of the following corollary is also left as an exercise.

Corollary *If* $\angle ABC \cong \angle A'B'C'$ *in such a way that* B *corresponds to* B', D *is a point of the interior of* $\angle ABC$, *and* D' *is a point of the plane* $A'B'C'$ *on the same side of* $A'B'$ *as* C' *such that* $\angle ABD \cong \angle A'P'D'$, *then* $B'D'|$ *is interior to* $\angle A'B'C'$.

EXERCISES

1. Prove Proposition 4.

2. Prove the corollary of Proposition 4.

3. Prove the corollary of Proposition 3.

4. In the proof of Proposition 2, prove that O' cannot be in (AO).

5. In the proof of Proposition 1, prove that $[AA']$ contains all of the points of AB which are interior to σ'.

6. Prove a.s.a. congruence. Specifically, prove that if $\triangle ABC$ and $\triangle A'B'C'$ are such that $(AB) \cong (A'B')$, $\angle BAC \cong \angle B'A'C'$, and $\angle CBA \cong \angle C'B'A'$ with A and B corresponding to A' and B', respectively, then

$$\triangle ABC \cong \triangle A'B'C'.$$

7. Let (AB), (CD), and (EF) be any three segments.
   a) Formulate a precise definition for the statement "any two of these segments are greater than the third." Begin by defining what it means for one segment to be greater than another segment.
   b) State and prove the proposition from Euclid, quoted at the beginning of this section, in terms of your definition.

8. Define the *bisector* of $\angle ABC$. Prove that if $\angle ABC$ is not a straight angle, then there is one and only one bisector of $\angle ABC$.

9. Let G be the set of congruencies of a plane (into itself). Prove that G is a group with function composition as its operation. Prove or disprove each of the following:
   a) G is transitive with respect to the set of lines.
   b) The plane is homogeneous.
   c) If A, B, C are noncollinear points, and if A', B', and C' are noncollinear points such that $(A, B) \cong (A', B')$, $(A, C) \cong (A', C')$, and $(B, C) \cong (B', C')$, then there is a unique member f of G such that $f(A) = A'$, $f(B) = B'$, and $f(C) = C'$.

10. Define a total ordering on the set of congruence classes of angles, and derive some of the properties of this ordering.

7.2 PERPENDICULAR LINES

Thus far no parallel axiom has been introduced. In point of fact, by using Axioms 1 to 12 with slight modifications (and one or more additional axioms), one can define the basic non-Euclidean geometries. Although we will not pursue this line, the possible use of Axioms 1 through 12 in setting up a non-Euclidean geometry is one of the advantages of this set of axioms. Almost all of the propositions we have proved thus far are true in non-Euclidean geometry as well as Euclidean geometry (but not all are true in all non-Euclidean geometries, as we will see in the next section).

Closely related to the notion of *parallelism* is the idea of *perpendicularity*. The reader might recall such Euclidean propositions as this one: Two lines perpendicular to the same line are parallel. Although we have no parallel axiom, our axioms already imply something about *perpendicularity* (a concept which we now define and investigate); and, hence, the axioms tell us something about parallel lines as well.

Definition 3 *An angle congruent to either of its supplementary angles in such a way that the vertex of the angle corresponds to itself is called a **right angle**. The two sides of a right angle are said to be **perpendicular** as are the two lines of which the sides are a part.*

The following is a trivial corollary of Definition 3.

Corollary *Both supplementary angles and the vertical angle of a right angle are right angles.*

Proposition 5 *If P is any point and AB is any line, then there is one and only one line which contains P and is perpendicular to AB in any plane which contains A, B, and P.* (This proposition is found in Euclid as Propositions 11, 12, and 14 of Book One. Our construction in the first paragraph of the proof is essentially the same as that which Euclid uses to prove his Proposition 11.)

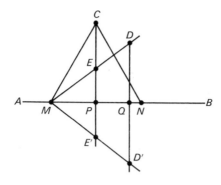

Fig. 7.7

Proof Assume that P is a point of AB. Choose points M and N of AB on opposite sides of P such that $(M, P) \cong (P, N)$. Let C be a point such that $\triangle MNC$ is equilateral. Now $(C, P) \cong (C, P)$, $(C, N) \cong (C, M)$, and $(N, P) \cong (M, P)$; therefore $\angle CPM \cong \angle CPN$ with P corresponding to P. Consequently, CP is perpendicular to AB. We now show that CP is the only line in ABC which contains P and is perpendicular to AB.

Suppose some line DP contains P and is perpendicular to AB. Then D is either on the same side of CP as M or on the same side as N; we can assume that D is on the side of N (Fig. 7.7). Since D and M are then on opposite sides of CP, (MD) meets CP in some point E. Let E' be the point of CP on the opposite side of P from E such that $(P, E) \cong (P, E')$. Let D' be the point of $ME'|$ such that $(M, D) \cong (M, D')$. Since $(E, P) \cong (E', P)$, $(M, P) \cong (M, P)$, and $\angle MPE \cong \angle MPE'$, we have $\triangle MPE \cong \triangle MPE'$; hence $(M, E) \cong (M, E')$. Since E is in (MD) and E' is in $MD'|$, by Proposition 15 of Chapter 6, we have E' is in (MD'). Therefore, (DD') meets AB in some point Q different from P. This latter statement follows from the fact that D and D' can now be concluded to be on the same side of $EE' = CP$ (the side opposite M), but they are on opposite sides of $MN = AB$.

Since $\triangle MPE \cong \triangle MPE'$, we have $\angle EMP \cong \angle E'MP$. Consequently, since $(M, D) \cong (M, D')$ and $(M, N) \cong (M, N)$, we have $(D, N) \cong (D', N)$. Since DP is perpendicular to AB by assumption (that is, $\angle DPN \cong \angle DPM$), it follows that $(D, M) \cong (D, N)$. Similarly, $(D', M) \cong (D', N)$. Hence we have that (D, M), (D, N), (D', N), and (D', M) are all congruent. Therefore (D, D'), (D', N), and (D, N) are respectively congruent to (D, D'), (D, M), and (D', M). Thus, $\angle ND'Q \cong \angle MD'Q$; hence $(M, Q) \cong (N, Q)$, but $P \neq Q$. This, however, is a contradiction of Proposition 2. Therefore, PC is the only line which contains P and is perpendicular to AB.

Assume now that P is not in AB. Let P' be a point on the opposite side of AB (in the plane ABP) such that $\angle P'AB \cong \angle PAB$ and $(P, A) \cong (P', A)$. Such a point P' can be shown to exist by the use of Proposition 4. The proofs that P' exists and that PP' is perpendicular to AB are left as exercises.

As part of the proof of Proposition 5, it was shown that if *D* is not on *CP*, the line which contains *P* and is perpendicular to *AB*, then (*D*, *M*) is not congruent to (*D*, *N*). This gives rise to the following corollary.

Corollary 1 *If A and B are distinct points of some plane π, then the unique line in π, which is perpendicular to AB through O, the midpoint of (AB), contains all points X of π such that (A, X) ≅ (B, X). We call this line the* **perpendicular bisector** *of (AB) in π.*

Corollary 2 *Any two right angles are congruent to each other.*

The proof of this corollary is left as an exercise.

In previous propositions, we had to prove explicitly or assume that congruencies took vertices of angles to vertices. Actually, this stemmed from the fact that we did not know if congruencies had to take collinear points to collinear points. They do, as we see from the next proposition.

Proposition 6 *A set of three noncollinear points cannot be congruent to a set of three collinear points.*

Proof Suppose *A*, *B*, and *C* are any three noncollinear points, and suppose *P*, *Q*, and *R* are three collinear points (Fig. 7.8). If (*A*, *B*), (*B*, *C*), and (*C*, *A*) were respectively congruent to (*P*, *Q*), (*Q*, *R*), and (*R*, *P*), then there would be a point *D* on *CA* such that (*A*, *B*) and (*C*, *B*) were congruent to (*A*, *D*) and (*C*, *D*), respectively. Let *M* be the midpoint of (*BD*). Then △*AMB* ≅ △*AMD*. Therefore, ∠*AMD* ≅ ∠*AMB*; hence these angles are right angles, and *AM* is perpendicular to *BD*. But since (*B*, *C*) ≅ (*D*, *C*), it also follows that *CM* is perpendicular to *BD*. Since there is only one perpendicular (in *BAC*) to *BD* at *M*, it must be that *A*, *M*, and *C* are collinear. Thus, *AC* = *MD*, and *A*, *B*, and *C* are collinear, contradicting the assumption that they are noncollinear.

Corollary 1 *A congruency between two angles which are not straight angles, or between two triangles, takes vertices to vertices.*

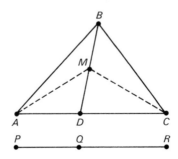

Fig. 7.8

Corollary 2 *Any congruency of a plane onto itself is a collineation. Any congruency between two planes is an isomorphism.*

The proofs of the corollaries are left as exercises.

Not only does a congruency preserve collinearity, but it also preserves *betweenness* as we see from the next proposition.

Proposition 7 *Suppose f is a congruency from S onto T and A, B, and C are points of S such that B is in (AC). Then $f(B)$ is in $\big(f(A)f(C)\big)$.*

Proof Since A, B, and C are collinear, $f(A)$, $f(B)$, and $f(C)$ are also collinear. Therefore we have either

$$f(B) \quad \text{in} \quad /f(A)f(C) \qquad \text{or} \qquad f(B) \quad \text{in} \quad /f(C)f(A).$$

Now

$$(A, C) \cong \big(f(A), f(C)\big), \qquad (A, B) \cong \big(f(A), f(B)\big),$$

and

$$(B, C) \cong \big(f(B), f(C)\big).$$

If $f(B)$ is in $/f(A)f(C)$, let B' be that point of $f(A)f(C)/$ such that $(A, B) \cong (f(A), B')$. Then B' and $f(B)$ are both points of $f(C)f(A)/$ such that (B, C) is congruent to both $\big(f(C), B'\big)$ and $\big(f(C), f(C)\big)$, contradicting Axiom 7. If $f(B)$ is in $f(A)f(C)/$, then by Proposition 15 of Chapter 6, $f(B)$ is in $\big(f(A)f(C)\big)$.

Corollary *If f is a congruency of the plane π into itself, and if A and B are distinct points of π, then*

$$f(AB) = f(A)f(B),$$

$$f\big((AB)\big) = \big(f(A)f(B)\big),$$

$$f\big([AB]\big) = \big[f(A)f(B)\big],$$

$$f(/AB) = /f(A)f(B),$$

$$f(AB/) = f(A)f(B)/.$$

The proof is left as an exercise.

We include the statement of the next proposition because it is needed for certain constructions. We omit the proof of this proposition because of its length; the reader can reconstruct the proof from the original paper of Veblen if he wishes.

Proposition 8 *A line which contains a point interior to a circle (and lies in the plane of the circle) contains precisely two points of the circle.*

EXERCISES

1. At the end of the proof of Proposition 5, (a) prove the existence of P', and (b) prove that PP' is perpendicular to AB.

2. Prove Corollary 2 of Proposition 5.

3. Prove the corollaries to Proposition 6.

4. Prove or disprove the following: Proposition 6 is equivalent to having any congruency between two triangles take vertices to vertices.

5. Prove the existence of the point D in the proof of Proposition 6.

6. Prove the corollary of Proposition 7.

7. Let π be any plane. We define a subset U of π to be *open* if, given any point A of U, there is a circle σ such that the interior of σ lies entirely in U.

   a) Prove that π and the empty set are open.
   b) Prove that the union of any family of open sets and the intersection of any two open sets is open.
   c) Prove that any set which is open in the above sense is open in the sense of Definition 3 of Chapter 6. Is the converse true?

8. Just using what we have so far, we can prove the first 28 propositions of Euclid's Book One. We now give some of these propositions as Euclid stated them. Restate each proposition, if necessary, in terminology consistent with that we have been using, and then prove the proposition. Assume everything takes place in some one plane. The proposition number from Euclid is given in parentheses; in many instances, the proof from Euclid will work.

   a) (2) From a given point to construct a segment equal to a given segment.
   b) (13) The angles which one straight line makes with another straight line on one side of it (ray) are either two right angles, or are together equal to two right angles.
   c) (16) If one side of a triangle be produced, the exterior angle shall be greater than either of the interior opposite angles.
   d) (17) Any two angles of a triangle are together less than two right angles.
   e) (18) If one side of a triangle be greater than a second side, the angle opposite the first side shall be greater than the angle opposite the second.
   f) (19) If one angle of a triangle be greater than a second angle, the side opposite the first angle shall be greater than the side opposite the second.
   g) (20) Any two sides of a triangle are greater than the third side.
   h) (27) If a straight line meeting two other straight lines makes alternate angles equal to one another, then the two straight lines shall be parallel.

7.3 PARALLEL LINES

As was pointed out in the previous section, the axioms we already have imply something about parallel lines. Before assuming a Euclidean parallel axiom, we will look at some of the things we have even without such an axiom. For the sake of an orderly discussion, we introduce some new terminology.

We will assume that everything takes place in some one plane.

Definition 4 By an **oriented angle** $\angle ABC$, we mean $BA/ \cup BC/ \cup \{B\}$, *together with either the interior or exterior of $\angle ABC$, but not both. Generally, $\angle ABC$ will be used to indicate the oriented angle*

$$\angle ABC \cup (\text{the interior of } \angle ABC),$$

or when we wish to make a statement which is applicable to both oriented angles ABC. We may also denote

$$\angle ABC \cup (\text{exterior of } \angle ABC)$$

*by $c(\angle ABC)$. We shall say that $\angle ABC$ and $c(\angle ABC)$ are **complementary**.*

Definition 4 enables us to discuss angles larger than straight angles. Although the notation we have chosen is potentially confusing because of the possible ambiguity of $\angle ABC$, it should be clear, in any given context, to which angle we are referring. Henceforth, any angle will be assumed to be oriented. Note that the interior of $\angle ABC$ is the exterior of $c(\angle ABC)$.

Definition 5 We say that $\angle ABC > \angle A'B'C'$ if there is a point D in the interior of $\angle ABC$ such that $\angle ABD \cong \angle A'B'C'$ (see Fig. 7.9). In keeping with the meaning of $\geq$, $\angle ABC \geq \angle A'B'C'$ will indicate that

$$\angle ABC > \angle A'B'C' \qquad or \qquad \angle ABC \cong \angle A'B'C'.$$

*Consider $\triangle ABC$ and let D be a point of $/BA$ (Fig. 7.10). Then $\angle CBD$ is said to be an **exterior angle** of $\triangle ABC$. The angles BAC, ACB, and ABC are said to be **interior angles** of $\triangle ABC$; $\angle BAC$ and $\angle ACB$ are said to be **opposite** $\angle CBD$.*

*Let AB and CD be two lines and MN be a line which meets AB in E and CD in F (Fig. 7.11). The angles MEB, MEA, NFD, and CFN are said to be **exterior angles** of the **transversal** EF, while angles AEF, BEF, DFE, and EFC*

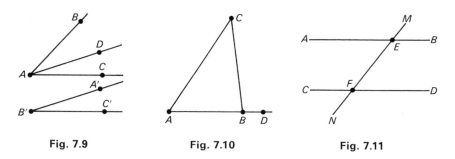

| Fig. 7.9 | Fig. 7.10 | Fig. 7.11 |

are **interior angles**. ∠ AEF and ∠ EFD are said to be **alternate angles**, as are ∠ BEF and ∠ EFC. Any two angles sharing a common vertex are said to be **adjacent**; two angles not sharing a common vertex are said to be **opposite**.

Although we have just introduced a good number of definitions, very few of them should be new to the reader since we have maintained the usual terminology of standard Euclidean plane geometry.

Proposition 9 (Euclid's Proposition 16, Book One) *An exterior angle of a triangle is greater than any interior angle opposite to it.*

Proof Consider △ABC and D a point of BA/. We must show ∠ DBC > ∠ BAC and ∠ DBC > ∠ ACB. We prove the latter and leave the proof of the former inequality as an exercise. Let E be the midpoint of (CB) and F be the point of /EA such that (A, E) ≅ (E, F), as shown in Fig. 7.12. Since (A, E) ≅ (E, F), (C, E) ≅ (B, E), and ∠ AEC ≅ ∠ BEF, it follows that ∠ ACB ≅ ∠ CBF. But F is interior to ∠ CBD; therefore

$$∠ CBD > ∠ CBF ≅ ∠ ACB.$$

Proposition 10 *In the situation of Fig. 7.11, if ∠ AEF ≅ ∠ EFD, then AB and CD have no point in common.*

Proof Suppose AB and CD share a common point G. Then G is either on the same side of EF as B, or as A. We prove it is not on the same side as A and leave the proof for B to the reader. Consider △FEG (Fig. 7.13). ∠ EFD is an exterior angle to △FEG and ∠ AEF is an interior angle opposite it. Then ∠ AEF > ∠ EFD, contradicting the assumption that these angles are congruent.

Proposition 11 (Euclid's Proposition 28, Book One) *If a straight line meeting two other straight lines is such that an exterior angle is congruent to the opposite interior angle on the same side of the transversal, or if the sum of the interior angles on the same side of the transversal is equal to a straight angle, then the two lines cut by the transversal do not meet.*

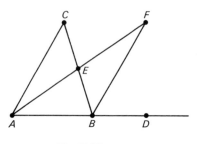

Fig. 7.12

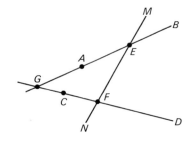

Fig. 7.13

Before proving this proposition, it is necessary to clarify what is meant by "the sum of two angles being equal to a straight angle."

Definition 6 *We say that the **sum** of ∠ABC and ∠A'B'C' is ∠MON if there is a point D in the interior of ∠MON such that ∠MOD ≅ ∠ABC and ∠NOD ≅ ∠A'B'C'. We designate that ∠MON is the sum of ∠ABC and ∠A'B'C' by writing ∠ABC + ∠A'B'C' = ∠MON.*

Restatement of Proposition 11 (Consider Fig. 7.11.) *If ∠MEB ≅ ∠EFD, or if ∠BEF + ∠EFD is equal to a straight angle, then AB and CD do not share a common point.*

Proof Suppose first that ∠MEB ≅ ∠EFD. Then ∠AEF ≅ ∠MEB (Proposition 23 of Chapter 6), and hence, ∠AEF ≅ ∠EFD. Therefore, by Proposition 10, AB and CD do not meet.

Suppose now that ∠BEF + ∠EFD is a straight angle. Then ∠BEF + ∠AEF is a straight angle. It therefore follows that ∠EFD ≅ ∠AEF (the proof is left to the reader). Therefore AB and CD do not meet.

Definition 7 *Two lines L and L' are **parallel** if L = L', or if L and L' share no points in common.*

Clearly, if L is parallel to L', then L' is parallel to L. We now prove half of what is generally taken as the parallel postulate for Euclidean geometry.

Proposition 12 *Let L be any line and P be any point. Then there is at least one line which contains P and is parallel to L.*

Proof If P is in L, then L is parallel to L. Assume P is not in L. Let L" be the unique line which contains P and is perpendicular to L, and let L' be the line which contains P and is perpendicular to L". It follows at once from Proposition 11 that L' and L are parallel.

Since one form of non-Euclidean geometry assumes that there are no lines parallel to a given line through a point outside the line, it is now clear that Axioms 1 through 12 cannot apply to all non-Euclidean geometries. This defect can be remedied by certain modifications of Axioms 1 to 12, but we will not pursue this point. In Euclidean geometry, of course, we want only one parallel. We therefore adopt the following axiom.

Axiom 13 *Given any point P and any line L, there is at most one line which contains P and is parallel to L.*

Combining Axiom 13 and Proposition 12 we obtain the usual statement of the Euclidean parallel postulate.

Proposition 13 *Given any point P and any line L, there is precisely one line which contains P and is parallel to L.*

Euclid did not state his parallel postulate in the form of Proposition 13. Instead, Euclid said the following:

Euclid's Fifth Postulate *If two straight lines in a plane meet another straight line in the plane so that the sum of the interior angles on the same side of the latter straight line is less than two right angles, then the two straight lines will meet on the same side of the latter straight line.*

We now prove the equivalence of Axiom 13 and Euclid's Fifth Postulate, given Axioms 1 through 12.

Proposition 14 *Assuming Axioms 1 through 12, Euclid's Fifth Postulate is satisfied if and only if Axiom 13 is satisfied.*

Proof First, assume Axiom 13 is satisfied. Let the situation be as in Fig. 7.14 with $\angle BEF + \angle DFE$ less than two right angles (that is, a straight angle). Let OP be the unique line which contains E and is parallel to CD. It can be proved (the proof is left as an exercise) that there are (unique) lines $O'P'$ and $O''P''$ such that $\angle O'EC \cong \angle EFD$ and $\angle P''EC + \angle EFD$ is a straight angle. Then $O'P'$ and $O''P''$ are both parallel to CD and contain E; therefore, $OP = O'P' = O''P''$. Since $\angle FEP > \angle BEF$, $EB/$ lies on the same side of OP as CD, and $EA/$ lies on the opposite side of OP from CD. Therefore, AB meets CD in a point of $EB/$; hence Euclid's Fifth Postulate is satisfied.

Now assume Euclid's Fifth Postulate. Let CD be any line with E a point not on CD (Fig. 7.14 may be used). Let $O''P''$ be the unique line containing E such that $\angle P'EF + \angle EFD$ is a straight angle. Then $O''P''$ is parallel to CD. Suppose $O'P'$ is any other line containing E. Then since

$$(\angle O'EF + \angle CFE) + (\angle P'EF + \angle DFE)$$

is two straight angles, it follows that a straight angle is greater than either $\angle O'EF + \angle CFE$ or $\angle P'EF + \angle DFE$. In either case, $O'P'$ meets CD.

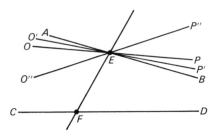

Fig. 7.14

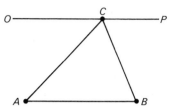

Fig. 7.15

Consequently, $O''P''$ is the only line which contains E and is parallel to CD; hence Axiom 13 applies.

Although we will not prove the equivalence, the following well-known theorem from plane geometry is also equivalent to Axiom 13.

Proposition 15 *The sum of the interior angles of a triangle is equal to a straight angle.*

Proof Consider $\triangle ABC$. Let OP be the unique line which contains C and is parallel to AC (Fig. 7.15). Then $\angle OCA \cong \angle CAP$; similarly, $\angle PCB \cong \angle CBA$. Since $\angle OCP$ is a straight angle, we have $\angle ACB + \angle CAB + \angle CBA$ is a straight angle.

EXERCISES

1. In the proof of Proposition 10, prove that G cannot be on the same side as B.

2. In the proof of Proposition 11, prove that if $\angle BEF + \angle EFD \cong \angle BEF + \angle AEF$, then $\angle EFD \cong \angle AEF$.

3. In the first part of the proof of Proposition 14, prove the existence of the line $O'P'$ and $O''P''$.

4. Does the linear space in Example 3 of Chapter 5 satisfy Axiom 13?

5. Explain carefully what the following statement means and why it is correct: The addition defined for angles in Definition 6 does not really tell us how to add two angles, but rather how to add two congruence classes of angles.

   a) Do the congruence classes of angles, with addition as defined in Definition 6, form a group? What group properties are lacking?

   b) Define "negative" angles and angles greater than two straight angles in such a way as to make the congruence classes of angles into a group.

6. Where was Axiom 13 needed in the proof of Proposition 15? Specifically, we know that there is a line OP which contains C and for which $\angle OCA \cong \angle CAB$. Moreover, we know that this line is parallel to AB. Where does the uniqueness of the parallel come in?

7. Prove that $\geq$, as defined in Definition 5, gives a total ordering of the congruence classes of angles (even though we are simply relating two angles in Definition 5). Prove several statements relating this total ordering with the addition of congruence classes of angles.

8. Prove that $\angle ABC \cong c(\angle ABC)$ if and only if $\angle ABC$ is a straight angle.

7.4 DISTANCE AND LENGTH

In this section we will develop a theory of distance measurement in the structure defined by Axioms 1 through 13. To be certain that distance has the properties we want it to have, we will have to introduce two final axioms.

Given two points A and B, it makes sense to define the distance between A and B to be the length of $[AB]$; we are then left with the problem of defining the length of $[AB]$. To define the length of $[AB]$, it suffices to define the length of any interval congruent to $[AB]$, since we certainly want congruent intervals to have the same length. Therefore, it suffices to assign a length to every interval one of whose end points lies on a ray $CD/$ and the other end point of which is C. Then we could assign to $[AB]$ the length of $[CP]$, where P is the unique point of $CD/$ such that $(A, B) \cong (C, P)$.

We expect the length of a (nondegenerate) interval to be given by some positive real number. This positive real number tells us how many *unit measures* of length are in the interval. For example, if we say that the length of a wall is twelve feet, then we mean that if twelve rulers, each exactly as long as some standard foot ruler, were laid end to end starting at one end of the room, then the end of the last ruler would touch the other end of the room. If some piece of string is $\frac{1}{2}$ meter in length, this means that if one end of the string were placed at one end of the standard meter rule and the string is stretched along the rule, then the other end of the string would be at the midpoint of the standard meter.

The standard unit length is generally arrived at rather arbitrarily. There is nothing intrinsically good about the mile, meter, foot, or inch. Authorities just decided to adopt these as unit measurements (along with others) and to express other lengths in multiples of these units.

Let $CD/$ be any ray. If we are to measure lengths of intervals with C as one end point and a point of $CD/$ as the other end point, we must first have a unit interval. We therefore choose any point U of $CD/$ and define the *length* of $[CU]$ to be 1. We will find the lengths of other intervals by determining what multiples of $[CU]$ they are. We now define what we mean by an *integral multiple* of any interval in $CD/ \cup \{C\}$.

Definition 8 *The line CD can be totally ordered as in Section 5.4. We lose no generality in assuming that $C < D$. Then*

$$CD/ = \{X \mid C < X\}.$$

$$Q = Q_1 \quad Q_2 \quad Q_3 \quad \cdots \quad Q_{n-1} \quad Q_n$$

$[PQ_n] = n[PQ]$ **Fig. 7.16**

Suppose P and Q are two points of $CD/ \cup \{C\}$ such that $P < Q$ and n is a positive integer. We define $n[PQ]$ as follows: Let $Q_1 = Q$. Let Q_2 be the point of $/Q_1P$ such that $(P, Q) \cong (Q_1, Q_2)$, and Q_3 be the point of $/Q_2Q_1$ such that $(P, Q) \cong (Q_2, Q_3)$. Continue in like fashion (Fig. 7.16) until Q_n is picked in $/Q_{n-1}Q_{n-2}$ such that $(P, Q) \cong (Q_{n-1}, Q_n)$. It is $[PQ_n]$ that we define to be $n[PQ]$.

Note that from the assumption that $P < Q$ we have

$$P < Q = Q_1 < Q_2 < \cdots < Q_n.$$

Also $1[PQ] = [PQ]$. We are now ready to begin assigning lengths to certain intervals.

Definition 9 *If O is any point of $CD/$ such that $[CO] = n[CU]$ for some positive integer n, then we define the **length** of $[CO]$ to be n. If O is any point of $CD/$ such that $n[CU] = m[CO]$, where m and n are positive integers and n and m have no common divisors, we define the **length** of $[CO]$ to be n/m. If the length of an interval $[CO]$ is r, then we set the length of any interval congruent to $[CO]$ equal to r. Denote the length of $[AB]$ by $L([AB])$.*

Definition 9 gives lengths for intervals which are rather nicely related to the unit interval $[CU]$, but we are still faced with assigning lengths to intervals not "nicely related" to $[CU]$. Moreover, we must be sure that a length is assigned to every interval. In particular, suppose there is a point O of $CD/$ such that O is not in $n[CU]$ for any positive integer n. Intuitively, if a point O lies in $n[CU]$, then $L([CO]) \leq n$; and if $L([CO]) \leq n$, then O must lie in $n[CU]$. Therefore, if O is not in $n([CU])$ for any positive integer n, then $L([CO])$ will not be assigned, or will be infinite. We may be willing to accept infinite lines, but an interval of infinite length is something else again. We will, in fact, need another axiom to ensure that any interval has finite length. Consider the following example.

Example 1 Let R and R' both be the set of nonnegative real numbers, except the elements of R will be considered distinct from the elements of R'. If you wish, consider the elements of R to be red and the elements of R' to be blue. Then $R \cap R' = \varnothing$. We totally order $R \cup R'$ as follows: Any element of R will be less than or equal to any element of R'. If s and t are both elements of either R or R', the usual ordering of R and R' will apply. The reader should confirm that $\leq$ then gives a total ordering of $R \cup R'$. Were $CD/ \cup \{C\}$ to look like $R \cup R'$, with C and U corresponding to 0 and 1, respectively, in

R, then no multiple of $[CU]$ would ever contain a point of $CD/$ corresponding to a point of R'.

To avoid the possibility of the situation illustrated in Example 1, we accept the following axiom.

Axiom 14 *Given the ray $CD/$ and the points U and O in $CD/$, there is a positive integer n such that O is contained in $n[CU]$.*

Axiom 14 is the so-called *Archimedean Axiom*.

We are now in a position to measure the "irrational" intervals of $CD/ \cup \{C\}$.

Definition 10 *Let O be a point of $CD/$ such that no length has yet been assigned to $[CO]$. We will call any point X of $CD/$, for which a length has been assigned to $[CX]$ by Definition 1, a **rational point** of $CD/$. Let*

$$S = \{X \mid X \text{ is a rational point of } CD/ \quad \text{and} \quad C < X < O\},$$
and
$$T = \{L([CX]) \mid X \text{ is in } S\}.$$

*By Axiom 14 there is a positive integer n such that O is in $n[CU]$. Therefore, we can conclude that n is an upper bound for T. Now every subset of the full set of nonnegative real numbers which has an upper bound has a least upper bound; therefore T has a least upper bound r. We define $L([CO])$, the **length** of $[CO]$ (as well as the length of any interval congruent to $[CO]$), to be r.*

We could prove the following proposition using only Axioms 1–14.

Proposition 16 *Define a function f from $CD/ \cup \{C\}$ into R', the set of nonnegative real numbers as follows: Set $f(C) = 0$. For each point X in $CD/$, set $f(X) = L([CX])$. Then:*

a) *f is one-one.*

b) *For any nonnegative rational number q, there is a point O of $CD/ \cup \{C\}$ such that $f(O) = q$.*

c) *If $O < P < Q$ in $CD/$, then $f(O) < f(P) < f(Q)$ in R'.*

We omit the proof of these statements because it would require too much work for the benefit that might be gained. That these statements are true, however, should be fairly obvious to the reader.

We do not know, however, if f is onto; that is, we do not know if, given any positive real number r, there is O in $CD/$ such that $L([CO]) = r$. The following axiom ensures that $CD/$ is *complete*. With the aid of this axiom it is possible to prove that f is also onto.

Axiom 15 *Every nonempty subset of CD/ which has an upper bound has a least upper bound.*

Proposition 17 *The function f as defined in Proposition 16 is onto R'.*

Proof Let r be any positive real number. Set

$$Q_r = \{q \mid q \text{ is a positive rational number less than or equal to } r\}.$$

For each q of Q_r, there is a point O_q of CD/ such that $f(O_q) = q$. Let n be any integer greater than r. Then the end point of $n[CU]$ on CD/ is an upper bound for

$$S_r = \{O_q \mid f(O_q) \text{ is in } Q_r\}.$$

By Axiom 15, then, S_r has a least upper bound O. Since f is order preserving, $f(O) = r$, the least upper bound for Q_r. Therefore each point of R' is the image of some point of CD/ $\cup \{C\}$; hence f is onto.

Because of the technical difficulties in rigorously proving many of the assertions about length, some texts resort to a "ruler postulate." The general form of a ruler postulate is: *There is a one-one order-preserving function from any line onto the full set of real numbers. If f is such a function from the line CD onto the real numbers R, then we can define the length of the interval* $[PQ]$ *in L to be* $|f(P) - f(Q)|$. This enables us to get a flying start on the process of defining length, but it offers little insight into the problems associated with defining length. Whether or not our approach, using propositions whose proofs are too involved to give in this text, is any better is a matter of opinion. It should be kept in mind, however, that most of our difficulties concern the foundations of the real numbers and the theory of functions of a real variable rather than geometry.

Thus far we have assigned a length to any interval, and we are also sure that, given any positive real number r, there is an interval of length r. It follows directly from the definition of length that two intervals have the same length if and only if they are congruent. We now define a *distance function*.

Definition 11 *Let A and B be any two points. Define D(A, B), the **distance between A and B**, to be 0 if A = B, and L([AB]) if A ≠ B.*

The following proposition expresses the basic properties of the distance function D.

Proposition 18
a) $D(A, B) = 0$ *if and only if* $A = B$.
b) $D(A, B)$ *is a nonnegative real number.*
c) $D(A, B) = D(B, A)$.
d) $D(A, B) + D(B, C) \geq D(A, C)$.

We leave the proof of this proposition as an exercise.

Although we will not do so, we could now develop all of Euclidean geometry from Axioms 1 through 15.

EXERCISES

1. Prove Proposition 18. To prove (a) you may use the fact that, given any positive real number r, there is a rational number q with $0 < q \leq r$. To prove (c), use Proposition 6.

2. Prove that equality holds in Proposition 18 (d) if and only if B is in $[AC]$.

3. Suppose $[AA_1] \cup [A_1A_2] \cup \ldots \cup [A_{n-1}A_n] \cup [A_nB]$ is a polygonal path joining A and B. Define the *length* of the polygonal path. Prove that the polygonal path of minimum length joining A and B is $[AB]$. That is, prove the Euclidean statement: The shortest distance between two points is a straight line.

4. Prove that the set $R \cup R'$ of Example 1 is totally ordered and has the properties claimed for it.

5. Define a set U in a plane to be *open* if, given any point P of U, there is a positive real number r such that $N(P, r)$ is a subset of U, where $N(P, r)$ is defined to be $\{X \mid D(P, X) < r\}$. Prove each of the following.

   a) The union of any family of open sets is open.
   b) The intersection of any two open sets is open.
   c) The plane and the empty set are open.
   d) $N(P, r)$ is an open set for any positive number r and point P. Suppose U is open in the sense of this exercise. Determine whether it is open in the sense of Definition 3 of Chapter 6; in the sense of Exercise 7 of Section 7.2. Is the definition of *open* in this section equivalent to either of the other definitions of *open*?

6. Let L be any line, and let L be totally ordered as in Section 5.4. Prove that there is a one-one order-preserving function f from L onto R, the set of real numbers.

7. Let ABC be any plane. Find a one-one and onto function f from ABC onto the usual coordinate plane R^2. Try to define f so that it will be an isomorphism.

General Index

INDEX OF SYMBOLS